Accidents:

Causes, Investigation and Prevention

James Thornhill

Published 2011 by arima publishing

www.arimapublishing.com

ISBN 978 1 84549 508 4

© Dr James Thornhill 2011

All rights reserved

This book is copyright. Subject to statutory exception and to provisions of relevant collective licensing agreements, no part of this publication may be reproduced, stored in a retrieval system, or transmitted in any form or by any means, without the prior written permission of the author.

Printed and bound in the United Kingdom

Typeset in Palatino Linotype, 12/12

This book is sold subject to the conditions that it shall not, by way of trade or otherwise, be lent, re-sold, hired out, or otherwise circulated without the publisher's prior consent in any form of binding or cover other than that in which it is published and without a similar condition including this condition being imposed on the subsequent purchaser.

This text is the copyright of Dr James Thornhill. It is presented as a book and forms part of a two day presentation and training on *Accidents: Causes, Investigation and Prevention*. As you can imagine, a great deal of time and effort (and, it is hoped, some expertise) has gone into this work. Any reasonable request to reproduce parts of this text for teaching purposes will normally be answered positively, but please do not use this information to design and deliver separate teaching. Thank you.

Swirl is an imprint of arima publishing.
arima publishing
ASK House, Northgate Avenue
Bury St Edmunds, Suffolk IP32 6BB
t: (+44) 01284 700321
www.arimapublishing.com

Contents

Introduction 9
Abbreviations 12
Terminology 13

Chapter 1: Understanding Accidents 14
1.1 Some historical background 14
1.2 Law and compensation 16
1.3 Herbert Heinrich 18
1.4 The modern development of HSE legislation 21
1.5 The strange attraction of the "unsafe act" 23
1.6 Statistics and the art of the obvious 26
1.7 Error theory: James Reason 29
1.8 Latent errors 34
1.9 What is an accident? 39

Chapter: 2 Psychological Explanations 43
2.1 Introducing Fred 45
2.2 Introducing ego-states 46
2.3 More on ego-states, and understanding Fred 52
2.4 Transactions, discounts, frame of reference, contracts 56
2.4.1 Transactions 56
2.4.2 Discounts 63
2.4.3 Frame of Reference 67
2.4.4 Contracts 69

Chapter 3: The Advantages of Good Accident Investigation . 75
3.1 Why should we investigate adverse events? 75
3.1.1 Reducing the likelihood and impact of future adverse events .. 76
3.1.2 Improving and correcting the SMS 76
3.1.3 Improving operating efficiency and increasing profits 77
3.1.4 Reducing danger to individuals 78
3.1.5 Maintaining good morale 79
3.1.6 Educating staff at all levels 81
3.1.7 Preventing the loss of good staff 81
3.1.8 Identifying violations of company procedures 82
3.1.9 Undertaking research 82
3.1.10 Complying with company or external rules 82
3.1.11 Preparing information for regulatory agencies 83
3.1.12 Preparing answers for the media and other groups 83
3.1.13 Providing protection against litigation 86
3.1.14 Making it look better 87
3.1.15 Finding someone to blame 87
3.2 The direct and indirect costs of accidents 87

Chapter 4: Investigating Accidents 90
4.1 The seven essential arts of the accident investigator 90
4.1.1 Which events do we investigate? 90
4.1.2 How far do we investigate? 93
4.1.3 Who undertakes the accident investigation? 93
4.1.4 Being well prepared 95

4.1.5 Having authority 96
4.1.6 The accident tool kit 96
4.1.7 Writing a report 96
4.2 What do you have in your investigation tool kit? 97
4.3 What you do and when you do it 99
4.3.1 First things first – safety and help 99
4.3.2 Once the scene is safe 100
1. Securing the area 101
2. Informing those who need to know 101
3. An initial tour of the area 102
4. Assessing how much time you have 102
5. Managing the team 102
6. Getting details of potential witnesses 103
7. Photographs and films 103
8. Notes, plans and diagrams 105
9. Preserving the evidence and taking samples 105
10. Putting together a basic idea of what happened 107
11. The investigation office 107
4.3.3 Preparing for the investigation 108
1. Updating the list of persons who might be involved 108
2. Briefing and liaison committee 108
3. Who else is going to be involved? 108
4. Securing electronic data 109
5. Securing documents and paper records 109
6. The standard accident notice 110

4.3.4 The investigation 111
4.3.5 Cleaning up 112

Chapter 5: Interviews and Witnesses 113
5.1 Interviews 113
5.1.1 The contract 113
5.1.2 Blame 114
5.1.3 Conducting an interview 116
5.1.4 How to ask questions 119
5.1.5 Interviews at the scene and follow-up interviews 123
5.1.6 Recording the interview 124
5.2 Witnesses 127
5.2.1 Getting witness information as soon as possible 128
5.2.2 Witness reliability 129
5.2.3 The hostile witness, the know-it-all and the expert 131
5.2.4 Special care 132
5.2.5 Problems with witnesses 133
5.3 Being on the receiving end: accident and criminal
     investigations 135

Chapter 6: Making Sense of the Evidence 140
6.1 The essential timeline and how sticky notes can make a
     difference 140
6.2 Using MORT 143

Contents

Chapter 7: The Accident Report..151
7.1 Contents of the report..151
7.1.1 Standard and individual accident reports..............................152
7.1.2 Writing the report..153
7.1.3 Using witness statements..162
7.1.4 Who should read the report?..163
1. The legal problem 163
2. The media and press..164
3. Management 164

Chapter 8: Preventing Accidents 166
8.1 The Safety Committee 166
8.1.1 Is it worth it? 166
8.1.2 Who should be on the committee?.......................................167
8.1.3 Contracts for an effective committee....................................169
8.1.4 The Safety Committee and Accident Investigation 172
8.2 The Accident Investigation and Prevention Committee..173
8.3 What difference does your Safety Management System make? 177
8.4 How to get everyone to report adverse events and why it is important...180
8.4.1 It is no good just telling them what to do 180
8.4.2 Bribery and corruption 182
8.5 Safety culture 183
8.5.1 Accident investigation is the key 188
8.5.2 The importance of community 190

8.6 The importance of feedback 191
8.7 Accident statistics and how to make them work for change
.... 194

Chapter 9: A Final Word 196
9.1 Don't go with the flow - keep swimming! 196
9.2 Saving lives and saving money 199

Resources 202
Contact 202
Enquiry check-list 203
Bibliography 210
Index 214
Footnotes 217

Introduction

This book is for anyone who is interested in accident theory and investigation, but in particular it has been designed to provide information for participants in my two day *Accidents: Causes, Investigation and Prevention* course. The course is referred to from now on as *ACIP*.

ACIP grew out of the realisation that there a lot of competent health and safety professionals who need a basic to mid-level introduction to accidents and their investigation. The existing books I am aware of are out of date, out of print, too detailed, or horribly expensive (though not necessarily all at once!)[1] so it seemed to me that there was room for a simple but up to the minute presentation at a reasonable cost. Some other works are much more detailed, some are much better at explaining the whole range of problems and solutions, and some are written by people with far more experience. But what I aim to present is a book which gives enough of the most interesting theory, and enough of the best practical advice, to enable health and safety managers in all kinds of industries to feel confident in undertaking accident investigations within their existing area of competence.

There are expert consultants who are called in when something goes badly wrong and results in fatalities and major disasters. And there are regulators and police officers who conduct accident investigations, for whom the current work might be useful as a refresher course, but it is not written with them primarily in mind. This book and the *ACIP* course are for people working in organisations, usually within the health and safety department, who know that investigating accidents is important, who may have to lead or be part of an accident investigation team, and who want some help in doing it.

Everyone brings their own approach and experience to understanding and solving problems. What I have added here, and this doesn't appear in any book on accidents that I am aware of, is some theory from Transactional Analysis. In order to understand why accidents happen and what we can do to prevent them, we need to understand why people behave in certain ways. Transactional Analysis provides a simple but powerful framework for this.

The taught course is in three modules, and these have been converted into chapters in the book. Module One (Chapters 1,2 and 3) provides an introduction and background. Chapter one is the historical background, and is, I hope, interesting in its own right. But it also serves to show how accident investigation doesn't exist in a vacuum but is part of a historical development and changes in the way societies view these things. In Chapter 2 we ask the questions why accidents happen and why people behave in ways which they know to be dangerous. In Chapter 3 we ask why we should investigate accidents and look at the advantages of undertaking accident investigation.

Module Two is concerned with the practical aspects of accident investigation. Chapter 4 sets out what an investigator needs to know. The practical necessities of what the accident investigator does, when, and how, are fully explored Of course each investigator will have their own methods but my hope is that there is enough detail here for every reader to find something useful and adapt it to their own use. Chapter 5 is on witnesses, interviews, and how to deal with them. This is an area many HSE managers will find difficult, and the use of TA should provide some new ways of understanding and solving problems. In Chapter 6 we look at various methods of

Introduction

assembling the evidence that has been gathered during the investigation and making sense of it. Then Chapter 7 goes on to discuss the writing of an accident report and how to produce a workable story telling those who need to know what happened.

Module Three is shorter than the others because all the essential information has already been put into place. Understanding accidents or investigating them might amuse a few individuals, but there is not much point unless accident investigation leads to loss control and the better management of risks. In fact, the best possible management of risks. How to achieve this is the content of this third module.

The table of contents is pretty thorough, and will provide a good overview of the content. It can also be used to find individual topics, so for this reason the index is fairly short. I have aimed to restrict index references to what I think the reader might be interested in and which is not obvious from the table of contents. At the end of the book there is also some information on resources; a bibliography; contact information and a useful Enquiry check-list.

I sincerely hope that this book, along with the *ACIP* course, will lead to very good risk control in many undertakings and that, as a result, lives will be saved, injuries and ill-health reduced, and unnecessary expenditure avoided. If so, the effort of writing it will be richly rewarded. Your experiences in using this book, and your comments and suggestions will always be welcome. And if you are able to tell me that by using this book you achieved the aim of a very good safety management system, I shall be delighted. Please do contact me at:

beechhillsafety@googlemail.com

Dr James Thornhill, Nottingham, July 2011

Abbreviations

ACIP	Accidents: Causes, Investigation and Prevention
AIPC	Accident Investigation and Prevention Committee
BBS	Behaviour Based Safety
HSE	(1) Health Safety and Environment
	(2) Health and Safety Executive. In this book the abbreviation HSE has the second meaning only in the bibliography and notes
JSA	Job Safety Analysis
LTA	Less Than Adequate
MORT	Management Oversight and Risk Tree
MSDS	Material Safety Data Sheet
SC	Safety Committee
SMS	Safety Management System
TA	Transactional Analysis

Terminology

This book uses the following terminology, which is explained in more detail in the text:

- an adverse event is an accident or an incident
- an accident results in loss, whereas an incident does not
- an incident is
    - either a near miss, which is an event that has the potential to cause loss but doesn't
    - or an undesired circumstance, which is a set of conditions or circumstances that have the potential to cause loss
- "loss" means injury, ill health, loss or damage.

Sometimes neat schemes have to take account of a less ordered reality. According to the definition above, this book should be called "Adverse Event Investigation" but would people know what that meant? So in fact I use the general term "accident investigation" as in the title, to mean all aspects of what we do when we investigate an adverse event.

Chapter 1: Understanding Accidents

1.1 Some historical background

Accidents have happened since the first humans, or their ancestors, tried to do something and got it wrong. Industrial accidents have happened since humans started setting up workshops for specialised jobs, and this too goes back a very long way. The code of Hammurabi, which is nearly 4,000 years old, provided for compensation in the case of accidents which were due to poor building practice. If a house collapsed and killed the owner then the builder should be put to death. Ancient societies used slaves in large scale and labour-intensive industries, for example in agriculture, mining, stone-cutting, ship-building and the production of pottery. There were inevitably many industrial accidents but they happened mainly to slaves, and the loss of slaves was just one of the financial problems that wealthy people had to cope with. Scientific analysis of cause and effect was a rare occurrence in the ancient world, but occasionally someone did stop and ask what they could do to make life safer, or at least cheaper. One of the earliest examples of protective equipment comes from Roman flour mills, where supervisors would bind muslin cloth round the mouths and noses of slaves grinding corn. This had two advantages. Firstly it stopped slaves from eating the corn, and secondly, they didn't seem to get so many lung problems from inhaling the flour dust.

The first modern industrial society developed in the United Kingdom, so the first modern industrial accidents happened there. At first there was no legislation to control the development of industry, and accidents were seen as things that "just happened". Pressurised boilers often exploded, mines sometimes collapsed, and workers were killed or maimed. There was no compensation, no sick pay, and no pension. Those who couldn't work didn't eat. Seeing the loss and suffering of individuals and families, some people started agitating for legislation to introduce basic safety measures, but measures of reform were opposed by most factory and mill owners. They said that any attempts to introduce health and safety regulations would cost too much money and bankrupt them. They pointed out that if they went out of business there would be no work, then everyone would starve, and the country would collapse into anarchy.

Despite these dire warnings the UK parliament did eventually pass some laws regulating the most dangerous activities, like mining. Standards governing the production of dangerous equipment such as pressurised boilers were also introduced. Legislation started to regulate the minimum age of children who could be employed, and established Sundays as a day off. And contrary to the awful predictions of the mill owners, things generally improved and the country flourished.

At this stage, there was no formal accident investigation, and there were no requirements to report accidents to anyone. But when accidents happened, people did talk about them, and they sometimes drew conclusions about what had gone wrong. An exploding boiler or a collapsing mine shaft not only killed people but also cost a lot of money. If a new boiler had to be built, for example, it made sense to ask if anything could be learned from what hadn't worked last time. By learning from mistakes, design could be improved and the new equipment might, with a bit of luck, work more reliably and kill fewer people. That meant less trouble and bigger profits.

1.2 Law and compensation

The regulation of industrial plants and premises continued, but attempts to introduce compensation for workers who were injured remained stuck in a legal jungle. In any society, the main function of law is to protect the rights and property of the wealthy and powerful since, after all, they are the ones who make the law and inflict its penalties. At this stage in industrial development, the laws of the land did a good job in this respect, and generally speaking judges agreed that the imposition of costly safety measures upon mill owners was not an acceptable option. Accidents were understood as an unfortunate but necessary part of the industrial regime.

However, large numbers of workers continued to be killed or maimed at work, and there were some attempts to claim compensation from companies. They did not succeed. The law provided a number of defences for the employer, the first of which was the doctrine of *volenti non fit injuria*. This said that anyone who agreed to do a job also agreed to accept the risks which were inherent in the job. This was understood as a contract: on the one side someone (the employer) offered employment and money, and on the other side someone else (the worker) agreed to do the job at the agreed rate of pay. The worker was understood to be freely volunteering to undertake the task, and in doing that he or she also agreed to accept all the risks attached to it. Clearly, if the worker thought it was too dangerous, they didn't have to enter into the contract and do the work. Workers therefore had no right at all to come along later complaining that the job wasn't safe and that they had suffered injury as a result. Such an argument was, of course, complete humbug. Workers took the employment, safe or otherwise, because they had little choice. If they didn't work, they and their families would starve. It was "his poverty, not his will, consented to the danger".[2]

In addition, if there was an accident and a worker was injured, it might well be that either he, or some fellow worker, had contributed to the disaster through "negligence". This was enshrined in the principles of *contributory negligence* and in the *doctrine of common employment*. Suppose that a worker (let's call him John) has been employed to stoke the boilers in a factory. For ten hours a day, six days a week, he shovels coal into the furnace. He is supposed to keep an eye on the pressure gauge but he is tired and hot, and sometimes hungry, and sometimes ill; and sometimes he forgets. A fellow worker is supposed to clean out the pipes every so often, but this week he has been ill and hasn't done the job properly. Suddenly, the boiler reaches critical pressure and bursts, sending steam and chunks of metal in all directions. John is burned and loses an arm. Since he cannot work, he also loses his job. This means that neither he nor his family has any income; and there is no National Health Service yet so he has to pay if he wants the services of a medical doctor. Can he ask the employer for help or for compensation? The law said not. In the first place he had accepted the risks (*volenti non fit injuria*); secondly if he had looked at the pressure gauge he could have avoided the accident (so he was guilty of *contributory negligence*); and thirdly, a fellow worker had also been partly responsible in not maintaining the equipment properly (*doctrine of common employment*). The employer was not to blame.

Thus it was that workers were injured in the most horrific situations but were unable to successfully claim compensation. But there was another and long-term effect, which was to enshrine the belief that any accident was always the fault of the worker. The "negligence" of the worker was a key legal concept, and until 1945 any worker who had contributed in even the smallest degree to an accident would be unable to claim compensation from the employer. In any legal argument or compensation claim it was therefore essential for the employer to prove the worker's contributory negligence in every case.

1.3 Herbert Heinrich

In 1931 Herbert William Heinrich published an important book which helped focus interest on accidents and their causes.[3] He was not the first to investigate accident causation, but he was very influential for a long time. Heinrich worked for an insurance company, and was therefore interested in the question of why accidents happen. Presumably his aim was to prevent accidents by understanding their causes, and thereby save the insurance company from having to make expensive pay-outs. He examined large numbers of accident reports from the 1920's, and concluded that 88% of accidents were caused by human failure – in other words, the worker was to blame. We should note that the reports that Heinrich read were compiled by supervisors. Of course supervisors were going to blame the workers; who else was there to blame? Certainly not the supervisors.

Heinrich also concluded that an accident was the result of a sequence of events or states, and he illustrated this by a diagram showing five dominoes standing on end. The fifth domino is the accident, and the other four are things that precede it. The central and most important domino stood for unsafe acts and unsafe conditions. Heinrich based this accident theory on a number of "axioms". These were statements which Heinrich believed to be self-evidently true. Actually, they were not based on research or statistical analysis, but were widely assumed to be.[4]

Paradoxically, one of the problems with Heinrich's work was that it invited company management to consider accidents and their costs. Surely, you will say, that was a good thing? And, yes, in the long run it was, but for quite a time it actually had the effect of stopping any renewed interest in accidents and their causes. This isn't entirely true because there were other researchers who came to very different conclusions, but for those who followed Heinrich's analysis it was clear that all that could be done was being done. At the same time accountants were becoming increasingly important in management, and their job was to work out how much it cost to produce whatever it was that the company was producing, and to see how much profit could be made from that product. So it made sense to include all the unavoidable costs in the calculations and to set the end price at a level to cover them and still make a profit. Accidents were often seen as an unavoidable cost, just as they had been for thousands of years. Accountants weren't calculating how many slaves they might lose this year, but they certainly did work out how many accidents there might be, based on past performance, and what the cost of that would be. Then they factored these costs into the balance sheets and forgot about them.

Accident prevention took a back seat. As long as the costs could be written off in the accounting system they could be simply absorbed. And it wasn't just the costs of accidents within the factory. There is the infamous example of the Ford motor company's calculations concerning the production of the Pinto motor car. What actually happened is disputed, and Ford has always denied that it did any more than estimate how many accidents the car would be involved in and what the costs to the company might be. It has also been claimed that the Pinto was no worse, in terms of safety and the number of accidents it was involved in, than similar cars being produced in America at the time. For some, though, the suspicion remains that Ford deliberately under-engineered the vehicle's safety in order to save costs. The Ford company, say some researchers, seems to have weighed the cost of improving the car's safety against the cost of being sued as a result of serious and fatal accidents, and to have decided not to invest in the additional production costs. It was cheaper to have the accidents or, in this case, to let the customer have the accidents. What does seem clear is that the production of the Pinto was rushed through in the face of serious competition from small car manufacturers. When pre-production crash tests showed that the car was liable to explode in rear-end collisions the assembly lines were ready to go, and it would have cost a lot of money to set them up anew. Engineers knew that a single piece of metal weighing one pound and costing $1 would have solved the problem, but the CEO of Ford, Lee Iacocca, had set an absolute limit of 2,000 pounds weight and $2,000 cost. The safety cap was rejected, and the car was put on the market without it. A conservative estimate is that 500 people died in crashes involving the Pinto, and most of these deaths were truly terrible.[5] Although there was evidence that the car was unsafe, it was pushed into production through a management system that seems to have been confused and fragmented. At the time, the risk of accidents was often understood as just another cost of production.

Vincoli refers to the concept of "blood priority" popular at the time, which meant, apparently, that if blood wasn't spilled then there wasn't any priority, or a budget, for accident investigation.[6]

1.4 The modern development of HSE legislation

What changed things was society's attitude. In fact this is usually what changes things, and we can take the building of nuclear power stations as an example. Until recently, many governments intended to invest heavily in nuclear power, but after the accidents at Fukushima, society's attitudes have changed dramatically. As a direct result of the disaster in Japan, Switzerland and Germany announced that they would phase out all nuclear power, and in Italy a proposal to revive nuclear power was voted down.[7]

In the 1980's a similar shift in cultural attitudes took place, and again this was the reaction to a series of accidents as a result of which lots of people said "enough is enough". The decade saw the worst ever industrial accident at Bhopal (1984); the world's worst ever single plane crash (Japan Airlines, 1985); the Challenger space shuttle disaster (1986) and, in the same year, the nuclear accident at Chernobyl[8]. People still wanted progress, it is true, but not at the cost of major accidents. Another point was that until quite modern times, although large numbers of workers might be killed in accidents, the causes were readily understood. An exploding boiler or a mine collapse might cause death and destruction, but the immediate causes were easy to see and therefore to accept. That might sound strange, but research shows that people's assessment of risk is rarely accurate. We tend to be more afraid of relatively unknown risks over which we have little control and to accept risks which are known to us and over which we have some control.[9] So, for example, there are many people who are frightened of flying on commercial flights, regarding this rare activity as especially risky (and particularly after a warning of potential terrorist action). What they ignore is the risk of staying at home and being struck by a car or run over by the proverbial bus. Statistically it is far safer to travel by air than to cross the local main road; but air travel is a relatively unknown and unusual risk, over which we have little control, so it seems more dangerous.

So a mine collapse doesn't bother us as much as what is going on in that chemical plant over there. Most of us don't understand chemical plants, or nuclear power stations, or large aeroplanes, and so we tend to overestimate the risks. But we do know that as production and transport methods have become more and more complex, they have become more dangerous. People overestimate the hazards and demand that the risks be controlled. We are used to cars by now, but large aeroplanes are still relatively infrequent experiences for most people. So the public demands the strictest safety measures to ensure that flying machines are safe. In fact, the number of safety measures may in the end make the aeroplanes less safe than they would otherwise be ... but that is another story.[10]

Finally, there is the realisation that accidents in certain kinds of process can lead to catastrophic loss. The events at Chernobyl are a good example. While it is not known how many people died in the long term from the effects of this nuclear accident, there is no doubt at all that people all over the world woke up to the realisation that it could have been a lot worse and that a similar event could cause tens of thousands of deaths. And add to this the realisation that events in one part of the world can lead to harm in another, and people begin to demand very safe systems of work. Accidents are no longer acceptable.

Having looked at some of the historical background to accidents and the way we react to them, we turn to the question of how we understand the causes of adverse events.

1.5 The strange attraction of the "unsafe act"

One of Heinrich's "axioms" was that 88% of accidents result from unsafe acts and 10% from unsafe conditions, hence the importance of the central domino. Well, in one sense this is obviously true. If you investigate an accident you will always find an unsafe act or condition as an immediate cause. But this is not actually very helpful, and, worse, it leads accident investigation in the direction of blaming the immediate operator and ignoring wider influences. If you go into an accident investigation looking for an unsafe act, you will inevitably find it, but if you stop there you won't find much that's useful.

> "Blaming individuals is ultimately fruitless and sustains the myth that accidents and cases of ill health are unavoidable when the opposite is true."[11]

An unsafe act is committed by a person, and so you have someone to blame. As Heinrich himself said "man failure is the heart of the problem and the methods of control must be directed toward man failure." Management agreed that it could reduce accidents by removing the central domino, in other words, unsafe acts. To do this they invested in the three "E's"; Engineering, Education and Enforcement. There is nothing wrong with these in their right place, but they focus narrowly on the worker at the point of production. Manuele has this to say:

> "For years many safety practitioners based their work on Heinrich's theorems, working very hard to overcome "man failure," believing with great certainty that 88% of accidents were primarily caused by unsafe acts of employees. How sad that we were so much in error."[12]

Remember also that Heinrich was looking at accidents that happened in the 1920's. That is 90 years ago, when operatives standing in front of and operating machines was a lot more common than it is now. Working practices were very different; machines were very different; and people's understanding of and attitudes to adverse events was very different too. It is arguable that the information that Heinrich used to arrive at his axiomatic conclusions is not very relevant to understanding our subject in the modern world.

So we need to ask, why has the concept of the unsafe act been so popular? There are a number of reasons. In the first place it is simple. Everyone can understand it and relate to it. And it is also very attractive because it gives us the impression that we can do something about it. Workers do unsafe acts, so all we have to do is engineer out the obvious problems, then train the workers so they know what to do, and then enforce the rules so that they do what they know what to do. This is relatively cheap, and it gives us control. Well, sort of.

But the main attraction of the unsafe act is more sinister. Who wrote those reports that so interested Mr Heinrich? The supervisors. And whom did they blame? The workers. But there was no blame, as you will have realised by now, for the supervisors. As for the supervisors of the supervisors, and the management who appointed the supervisors, and the owners and directors who appointed the management, they weren't even considered. It was clearly the chap at the sharp end who was to blame. The laws which stopped workers from claiming compensation did so by putting the blame firmly on the workers, and Heinrich merely continued that system. Concentrating on the unsafe act as the cause of 88% of accidents allowed managers and supervisors to evade nearly all responsibility.

If management is often keen to get itself off the hook by pointing to a guilty individual, it can also be true, as Sutherland and colleagues point out, that there may be considerable pressures on the individual to accept the blame.[13] A worker may believe that they will keep their job if they own up and admit their guilt. The offer of a small amount of compensation may seem very tempting if the alternative is to fight the system and, perhaps, spend years in litigation with no assured result. And few other employers are likely to want to take on someone who is involved in such litigation. So it may seem easier to take the blame, and a small injury compensation package, and draw a line under the whole business.

The concept of the "unsafe act" is important, and is widely used today in accident prevention programmes, particularly in Behaviour Based Safety. But it is not sufficient as an explanation for adverse event causation. In particular, the way we understand the causes of adverse events is important when it comes to knowing which accidents to investigate. If our view is that we investigate accidents for "the prevention of further similar casualties" then we have narrowed our scope very considerably. We might make the wrong decision here, by looking at what happened and thinking that it is so unlikely that this accident is going to happen again that we do not need to pay much attention to it. This however is to concentrate on the outcome of the accident, not its generation. If we look only at the outcome and the immediately preceding unsafe act, then we may miss the more serious problems which lie behind it. We might be right in assuming that this particular accident isn't going to happen again, but another accident might happen and be the result of the same underlying problems that caused the accident we didn't investigate. On the other hand, investigating every accident is going to take a lot of time.

1.6 Statistics and the art of the obvious

To mix metaphors horribly, Heinrich's statistics seem to have been a trap that led to a dead end. His figures do not seem to have been based on any statistical research, but we don't really know because the data and methods he used were not published.

Another of Heinrich's ideas that has had a great influence is the safety triangle. He introduced this in 1931[14] and it serves to illustrate his theory of accident causation: unsafe acts lead to minor injuries and, over time, to major injury. The safety triangle proposes that for every 300 unsafe acts there are 29 minor injuries and one major injury. The proposed ratio is roughly 1:30:300.

There are various forms of what is often called the "accident triangle", or even the "accident pyramid" showing a relationship between fatal or serious accidents at the top, and incidents which result in no damage or loss at the bottom. I found one in the Risk Management and Safety Manual of a large American city dating from 2000, which presents Heinrich's "Accident Triangle" as a proven statistical relationship, though how the ratios in the diagram (1:2:30) are meant to fit Heinrich's figures remains something of a mystery. In 1969 Frank Bird analysed nearly two million reported accidents and produced a triangle which gives a ratio of 1 major injury: 10 minor injuries: 30 damage only accidents: 600 incidents with no damage or injury. In 2003 ConocoPhillips Marine undertook similar research and concluded that for every fatal accident there are probably 300,000 at-risk behaviours.

Like others of Heinrich's assertions, the safety triangle is true in an obvious sense. We do of course hope that it is the case that in any standard industrial activity there will be far more incidents than serious accidents. And we would probably also expect that there would be very few, if any, fatal accidents. But to refine a concept like this any further presents some problems. Researchers have looked at the figures, and come up with various ratios between serious accidents and non-serious accidents, but how reliable are the figures in the first place? We can probably agree that it is extremely unlikely that a fatal accident in the workplace will go unnoticed and unrecorded. If a worker falls off a scaffold and is killed on impact with the ground then this is going to be recorded as a fatal accident and lots of people are going to know about it. So it might seem that our statistics for fatal accidents are going to be accurate and inclusive. Unfortunately, this isn't the case. What if the worker falls from the scaffold, is injured, taken to hospital, and then discharged. Suppose he is too ill to work and loses his job, and then dies, some months later. Is that going to be recorded as a work-based fatality? In some countries, if the medical services are on the ball, it may be; in many other situations it will not. Or think of another scenario: suppose that a worker finishes a long shift in difficult conditions, and leaves work feeling stressed, tired, and angry. Suppose that driving home from work this worker makes an error of judgement and that they are killed in a car crash. This has happened outside working hours, and so the fatality will not be related to workplace statistics but it may be that the situation in the workplace had a direct effect on the fatal event. Or what about workers who are exposed to hazardous substances like asbestos and who die of occupational disease many years after the exposure took place? Those deaths won't have been included in the annual fatality figures, but they are deaths nevertheless. And what about work activities which in some countries are simply excluded from the statistics, for example deaths or injuries arising out of the supply or use of flammable gas? If data for fatal accidents are unreliable, what hope is there for other accident statistics? We know very well that reporting levels are often poor and that many accidents, even quite serious ones, are neither reported nor recorded. So the statistics for these are likely to be unreliable too.

But even if the data for adverse events were accurate, there is another problem with the safety triangle as the accident to fatality ratio depends on the work being undertaken. The relationship between trivial accidents and fatalities for an office worker is likely to be different to that for an agricultural labourer or a construction worker. A bomb disposal officer or an electrician working on live high voltage transmission systems may not record many serious accidents but there might be a high probability of fatality.[15]

The safety triangle, therefore, points to an obvious probability: the more accidents there are in any given situation or profession, the more likely it is that fatalities will occur. This is not an accurate statistical relationship, but a simple observation that a larger number of accidents may indicate a more dangerous environment. Finally, the safety triangle refers to injuries; but an organisation could suffer catastrophic loss without injury or fatality. The explosion and fire at Buncefield in December 2005, for example, resulted in only 43 minor injuries but the estimated cost was a billion pounds Sterling.[16]

Relying on statistics or relationships between various kinds of accidents to assess the safety of our workplace is to use a very blunt instrument. Instead it is better to understand how accidents are generated, and that means looking at systems.

1.7 Error theory: James Reason

Professor James Reason is important because he has popularised some pretty technical stuff. What follows is a brief summary of Reason's theories, and anyone interested in more detail will enjoy reading his books.[17] They are well-written and often fascinating, but the language is probably rather complex for the average reader – he is after all a Professor of Psychology!

Reason points out that "error" is not some strange beast that inhabits an otherwise sensible mind. Failure and success are the results of the same kind of mental processes. Only the outcome defines whether we call it failure or not. Errors are defined by the thinking processes of the individual and can be divided into three basic types;

- Lapses
- Slips
- Mistakes

Slips and lapses are very common in all activities. We forget to do the right thing, or we trip or slip. When I checked out of a hotel recently, I left a suit hanging in the wardrobe. I forgot that it was there, so this was a **lapse**. Because I did not want to forget anything, the evening before I left I put my clothes near my case, on the floor of the room. But a suit was hanging in the wardrobe and I decided not to take it off the hanger and put in on the floor because it would get creased. Then in the morning, I thought "all my clothes are here" so I didn't check in the wardrobe. Later that day, in another hotel and indeed in another country, I started looking for the suit and *instantly* I knew that I had left it behind – I didn't need to check my suitcase to realise my mistake. So that information was in my head somewhere but it hadn't surfaced at the right time. Forgetting to check the wardrobe was a "lapse". In the end I got my suit back, but only after a lot of time and trouble.

My new hotel room was rather dark, and at one point I tripped over a pair of shoes I had left beside a chair. This is a **slip**. I simply didn't notice the obstruction, and tripped over it. No harm done in this case, and I moved my shoes into a cupboard (and didn't forget them!). These are everyday examples, but slips and lapses are a major cause of accidents and injuries. If you watch any worker doing a job, or a maintenance engineer disassembling and reassembling a piece of equipment, you will quickly observe many slips and lapses. I once conducted a Job Safety Analysis on a worker using a pillar drill, and in the space of an hour identified more than fifteen lapses and two slips. The man was an experienced worker, in fact he supervised others in using machinery like this, but his error rate was pretty high. His working methods were inefficient and risky.

In the case of slips and lapses, there is no intention behind the action. The third kind of error is different. A **mistake is** when I do have a plan, which I intend to use to fulfil my aim or intention, but the plan doesn't work for me and I do not achieve my aim. Perhaps the plan was not a good one, or perhaps the plan was OK but I failed to put it into operation properly. The wardrobe example will illustrate this. My aim was to arrive at the next destination with all my belongings, and I had a cunning plan. I put everything I needed to pack out where I could see it the night before, so that I could throw it in my suitcase and be ready to leave in good time. My plan was essentially a good one, but because of a lapse, it went wrong and I did not achieve my aim.

The Kegworth air disaster is a good example of a mistake. Shortly after take-off one of the engines went wrong and smoke started to pour into the cabin. The pilots shut down the right hand engine believing that this was faulty, when in fact it was the left hand engine that was in trouble. The pilots made a mistake: they intended to follow a good plan, that is to shut down the damaged engine and carry on to land with the good engine, but for various reasons they shut down the wrong engine and then interpreted further evidence to support their mistaken decision. This is also a very good example of the futility of blaming the operator for an unsafe act at the end of the process. Yes, the pilots were mistaken, but they had not had enough training on what was a new variation on the aircraft they were familiar with. Also, the vibration meters which would have told them which engine was the problem were very small and not easy to read; and in any case these indicators were notoriously unreliable and pilots routinely ignored them. So there are all sorts of questions about design and training, for a start.[18]

We now turn to consider two different classifications. These refer to breaking the rules deliberately, and differ from the three error types in two important respects. In the first place they are defined not only in terms of what happens in my head but exist in the context of a set of rules or accepted behaviours. In the second place they are deliberate acts. I know what I am doing and intend to do it.

A **violation** is when someone deliberately disobeys the rules, but not with the intention of doing harm. Sometimes we find routine violations in a system of work. There may be codes of conduct, rules, operating procedures and so on, but whoever should be adhering to these rules deliberately and habitually chooses not to do so. They do not do this maliciously, and intend to achieve a positive outcome, perhaps increased productivity. Why do they find ways round systems and devices which are designed to keep them safe? There are many potential reasons, and people are ingenious at finding new ways of doing things. Perhaps the safety devices make the job more difficult or slower, and the supervisor is demanding more productivity.

According to Reason, there are two main factors here. First, there is the "natural human tendency to take the path of least effort". In other words, if there is an easy way of doing it, that's the one we prefer, even if it isn't the safest. Secondly, if the rule or restriction doesn't seem important, and if there are not likely to be any horrible consequences if I by-pass it, then I am much more likely to violate the regulation.

Many work situations provide examples of the first factor, where people take the easy route. Removing safety guards from machines happens because people get fed up with working round them, especially if they are not well-designed. People don't wear appropriate PPE because it is uncomfortable or makes doing the job more difficult. It may be easier to take a short cut through the warehouse even though this is forbidden because of the danger posed by fork-lift trucks and other vehicles using the space. In all these cases, the path of least resistance will be much more attractive if there are no obvious sanctions. On the other hand, if I know that the supervisor or foreman is going to give me a hard time when he sees that I have removed the machine guard, then I am much less likely to do it. If "everyone" takes the short cut through the warehouse, then I am going to do it too, even though it is more dangerous than the longer route. But if there is a good chance that I will be stopped and cautioned, and if I know that if I am caught again then my bonus or my job is at risk, then the longer route becomes a much more attractive choice. This accepted behaviour is an important part of what is meant by a "safety culture".

Sabotage is also a case where someone deliberately disobeys the rules, but in this case they have the intention of causing harm. Sabotage is not an adverse event because it is not unexpected or unintended, but it should not be discounted for that reason. The possibility of sabotage is a real one if employees are angry or disgruntled, and a great deal of damage, and even injury and death, can result.

Finally, in accident investigation we should remember the "hindsight fallacy". Most workers and managers are not saboteurs and do not intend to cause injury, loss and damage. They probably have a plan, and their intention is to achieve some positive aim. Slips, lapses and mistakes do occur however, and this can result in an adverse event. At Chernobyl, the operators did not intend to cause a catastrophic failure. What they wanted to do was test some equipment, and to do this they shut down the protective systems.[19] Ironically, the engineers and electricians were testing a procedure which was designed to solve a possible safety problem. They knew that this problem could prove serious in the event of an emergency shut-down, so their intention was to find a way round it. Various alarms went off as they prepared for the experiment, but these were ignored as it was so important to complete the experiment. After the event, we all know what happened. With hindsight we are able to look back, and to us the sequence of events seems inevitable. But for the people there at the time, their aim was clear, and the possibility that the system would go into melt-down was, they thought, a remote one. They were wrong, but at the time they did not know that they were wrong.

We think like this all the time. We do a quick calculation and think "Oh yes, it'll be alright" and we carry on. In many cases, whatever it is we are doing turns out OK and so there we are – plan, implement, succeed. We don't look back to think what might have gone wrong. But sometimes things do wrong, and then we see things from a different perspective. That was really stupid, we might say; it was obvious that it wasn't going to work! But at the time that wasn't what we were thinking. We shall come back to why we do these things later, and look at some explanations.

1.8 Latent errors

Reason sets out the idea that many errors are inherent (or "latent") in the system. There are a number of diagrams which illustrate this idea, the latest version of which is called the Swiss Cheese model.[20] Reason starts with the idea that there is an accident trajectory, like an arrow. If the energy represented by the arrow continues on its way unhindered, then it will result in an accident, so in order to prevent this, we put defences in place.

In the earlier versions, these defences are represented by a single sheet, which has a hole in it, the "limited window of opportunity". If it so happens that when the arrow reaches the defences it passes through that window, then an accident will result. But where does the arrow come from? Well, it begins with three sources of latent errors; fallible decisions; line management deficiencies and psychological precursors of unsafe acts. Then it comes to "unsafe acts" which are active failures. The diagram shows one way in which the model might be portrayed.

Let us examine this model, taking as an example a factory worker who is using an electric drill. The drill is working at 240 volts and is plugged into the mains supply. But the protective sheath on the cable is worn and cracked, and live wires are exposed. The operator knows this, but carries on using the drill anyway. For a time all is well, but then the operator comes into contact with the live wires and receives an electric shock that causes a serious burn.

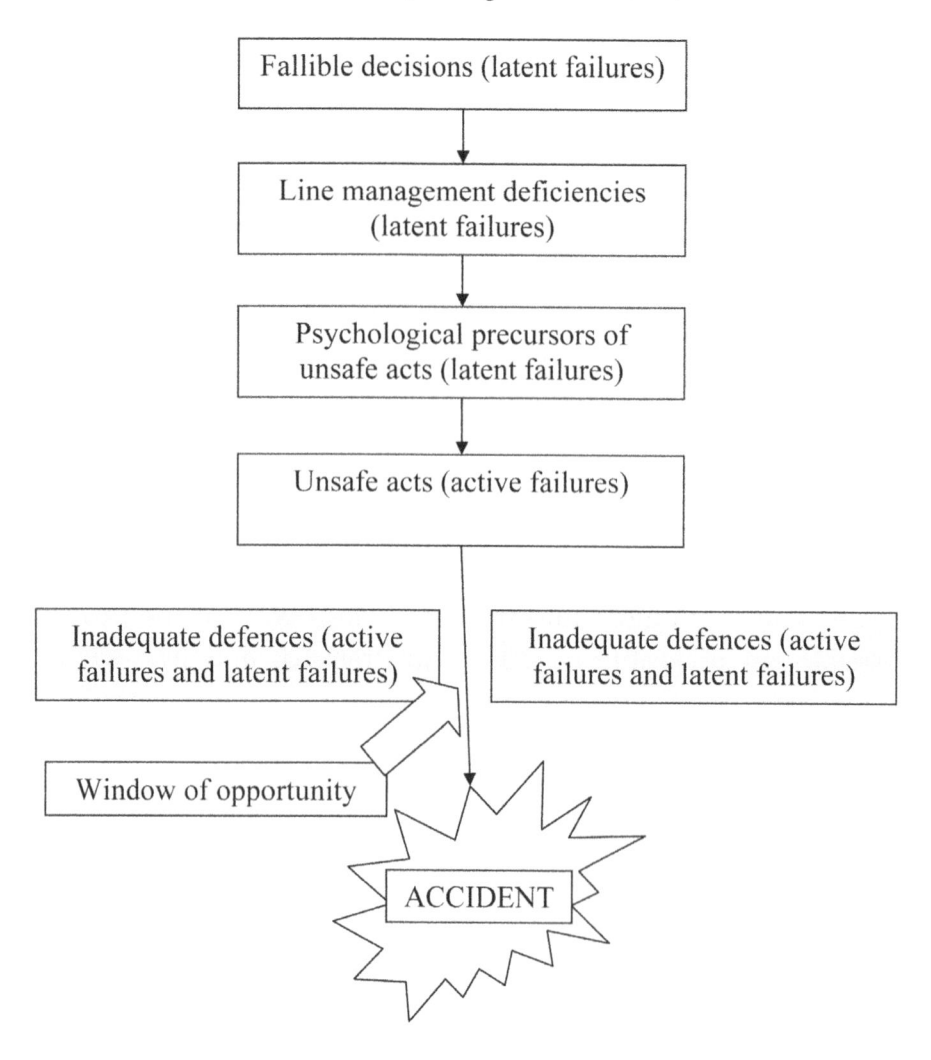

```
┌─────────────────────────────────┐
│ Fallible decisions (latent failures) │
└─────────────────────────────────┘
                  ↓
┌─────────────────────────────────┐
│   Line management deficiencies   │
│        (latent failures)         │
└─────────────────────────────────┘
                  ↓
┌─────────────────────────────────┐
│   Psychological precursors of    │
│   unsafe acts (latent failures)  │
└─────────────────────────────────┘
                  ↓
┌─────────────────────────────────┐
│   Unsafe acts (active failures)  │
└─────────────────────────────────┘
```

Inadequate defences (active failures and latent failures) | Inadequate defences (active failures and latent failures)

Window of opportunity

ACCIDENT

The accident is represented by the star at the bottom of the drawing. Above it are series of events, lined up a bit like Heinrich's dominoes. Immediately above is a panel representing inadequate defences, and in this panel there is a hole labelled "window of opportunity". The idea here is that we put defences in place to stop accidents from happening, but that no defence is totally effective. Any defence has the potential to fail, and if and when it fails, then there is an opportunity for an accident to happen. In our example, the insulation covering the

Chapter 1: Understanding Accidents

cable is a defence, a barrier against the potentially harmful energy of the electric current being used to power the drill. If the defence fails, in this case by becoming worn and broken, then an accident is much more likely to happen. The failure in the defence could be due to an active or a latent failure.

Knowing that the machine is faulty and there is high risk involved in using it, the operator continues nevertheless; this is an active failure, or in Heinrich's terminology, an unsafe act. Above that is a box representing the psychological precursors of the unsafe act. What was it in the history, background and psychology of the drill user that led them to carry on using the machine even when they knew it was not properly insulated? Moving up to the second panel down, "Line management deficiencies", we are invited to ask a vital question: "How was this accident allowed to happen?" How was it that the worker was using a defective drill? We might ask about the supervision of this worker, or about the procedures in place to ensure that faulty equipment is identified and removed from the workplace. We might ask about the worker's training and skill levels. We might ask about risk assessments, daily tool checks, operating procedures. And so on.

This is not the end of the matter, though. The panel furthest from the accident is entitled "Fallible decisions", and invites us to ask "What decision were made that contributed to this accident happening?" So, we might find out that management decided to cut costs by buying the cheapest equipment they could find; equipment which was perhaps not strong enough for the work. Latent errors might have been introduced into the system at this early stage. Or we might find that managers knew that this work would be carried out more safely if hand held drills were battery powered but, for reasons of economy, decided to carry on using the mains powered drills that they already had in the tool store. Or, we might discover that the management had decided that they would reduce costs by employing casual labour and reducing safety training to a minimum. This too could have contributed to the accident. And we might find that the company that designed and made the drill made a design mistake so that the cable was liable to fray and break easily at a particular point. And, again, so on.

Let us just pause for a moment and realise how very different this is to a simple understanding of accidents as the result of "operator error" or the dreaded "unsafe act". Using this more sophisticated model of accident causation, we can begin to see that many more factors are potentially important. The implications for our accident investigation are obvious. We are not going to be satisfied with a report which says that the worker used a drill when he shouldn't have. Yes, there was an unsafe act, but that is only one small part of the story. At different stages in different activities, like designing and manufacturing the drill, purchasing equipment, establishing operation and work procedures, recruitment, supervision, training, etc., decisions may be made and things done which will make an accident possible or more likely. All these things are called "latent errors". The Cambridge English Dictionary defines latent as something that is "present but needing particular conditions to become active, obvious or completely developed" and the Oxford Dictionary defines it as "(of a quality or state) existing but not yet developed or manifest; hidden or concealed".[21] Both these definitions are helpful; a latent error is something that is present in the system but which remains hidden until particular circumstances make it important.

One significant insight here is that latent errors may originate many miles away, and many years before an accident happens. Latent errors can sometimes be spotted and removed; and this is one of the reasons why we do accident investigations. But if they continue to exist unnoticed in the system then, given a certain set of circumstances, they may result in an adverse event. In our example, decisions made about the insulating material for the power lead on the drill may have been made years before the accident happened, and on the other side of the world. Purchasing decisions may have been made years earlier, by someone now no longer employed in the company, and in an office far removed from the scene of the accident.

We have certainly come a long way from the simple idea of the "unsafe act" as the cause of accidents. In the first place, we can no longer just blame the worker at the sharp end. The worker, tool user, machine operator, driver, pilot or nuclear power station control room supervisor is part of a system, not an independent and isolated actor. This has a number of very important results:

- Blaming the person immediately involved in the accident is seen to be just a way of covering up and ignoring the many different causes of accidents
- When an adverse event happens, it is because the system has failed
- It therefore follows that the prevention of adverse events depends on revising and correcting the system.

1.9 What is an accident?

The time has come to ask the basic question: what is an accident? There are many definitions, and most of them include the idea that an accident is unexpected or not designed. And most include the idea that the accident leads to loss of some kind. This can be damage to plant or equipment, loss of production, injury and harm to human beings or animals or the environment. A legal definition of an accident is "an unintended and unexpected occurrence which produces error or loss". Then it is important to add that an accident is the end product of a sequence of acts or events, in other words, accidents do not "just happen", even if they are unexpected. Then there is the idea that an accident is the unwanted transfer of energy to a target (which is a thing or a person). In order to prevent unwanted transfer of energy to targets we design barriers. An accident happens when there is (1) an unwanted transfer of energy to (2) a target and (3) the barriers to prevent that transfer are Less Than Adequate (LTA). So for example, we might put a machine guard on an electric table saw to prevent the transfer of mechanical energy from the blade to the operator's hands. If the guards are LTA the energy transfer may occur and injury may result. We can compare this with Reason's concept of the arrow of latent errors moving through a window of opportunity in defences.

But what if a volcano erupts and damages your house? Is that an accident? Perhaps we need to add something else to our definition, and conclude that an accident, at least for the purposes of accident investigation, is

- something that happens in the workplace
- which is unexpected and unintended
- at the end of a sequence of events
- and which leads to loss.

The word "incident" has different meaning for different people. Some companies have abolished the word "accident"; everything is an incident. This may sound good in the annual report, but will not be helpful for us, so we shall begin by defining an incident as the same as a "near miss", that is

- something that happens in the workplace
- which is unexpected and unintended
- at the end of a sequence of events
- and which does **not** lead to loss.

I agree with Manuele[22] when he asks how safety professionals can expect to be taken seriously if they can't agree on what terms are to be used to describe what they are dealing with. However I do not go along with Ferry[23] when he uses the term "accident" in the title of his book and then goes on to talk almost entirely about "mishaps". People are not going to stop using the words accident and incident – all we need to do is agree on what we mean. Some useful definitions are set out by the Health and Safety Executive in their excellent book, *Investigating accidents and incidents: A workbook for employers, unions, safety representatives and safety professionals* HSG245:

- an adverse event is an accident or an incident.
- an accident results in injury or ill health, whereas an incident does not.
- an incident is
    - a near miss, which is "an event that, while not causing harm, has the potential to cause injury or ill health"
    - an undesired circumstance, which is "a set of conditions or circumstances that have the potential to cause injury or ill health".[24]

I'm going to use this framework, as it is clear and agrees closely with my own understanding. So we shall use the term "adverse event" to include an accident or an incident, and include the "undesired circumstance" as part of the meaning of "incident". Curiously, though, this does not deal with financial or other kinds of loss. According to the Health and Safety Executive definitions, if a crane exceeds its working load and collapses onto a main road and there are, fortunately, no fatalities or ill health involved, this is not an accident. Most people however would certainly regard this as an accident and call it that. So I suggest that we revise the terminology as follows:

- an adverse event is an accident or an incident
- an accident results in loss, whereas an incident does not
- an incident is
  - either a near miss, which is an event that has the potential to cause loss but doesn't
  - or an undesired circumstance, which is a set of conditions or circumstances that have the potential to cause loss
- "loss" means injury, ill health, loss or damage.

And for our purposes we will understand that the adverse event happens in the workplace or in connection with work.

For convenience, I'll repeat here what I said earlier about loss: it is damage to plant or equipment, loss of production, injury and harm to human beings or animals or the environment. This is the best thinking I have at the moment on the meaning of the words we use in accident investigation.

Chapter: 2 Psychological Explanations

In our definition of adverse events we looked at what happens "out there". Something happens and impacts on us, so in attempting to avoid adverse events we look at what is going on in the world around us and do our best to change it by removing hazards, putting barriers in place, improving systems, etc. Now we need to look at another whole dimension to the system, and that is the human input. Why do we behave in the way we do? And what is even more puzzling, why do other people behave as they do, even when we can see that what they are doing is wrong, or odd, or silly?[25] Why did the worker in our example use the electric drill even though he knew that it was faulty and he could be injured if he carried on?

There is, I have found, a great deal of puzzlement in the business of safety management as to why people behave in ways which everyone knows are against their best interests. You can show a worker how to work safely, but then you find that they don't. How do you get someone to do what you want them to do, to change their behaviour so that they work safely? You can tell workers to wear ear muffs and hard hats, and explain to them that this is really a good thing for them, and you can show them grisly pictures and tell them horror stories about what happens to people who don't wear their PPE, but as soon as your back is turned, they take the hard hats and ear muffs off. And how do you get people to report incidents and minor accidents instead of covering them up? These are all very practical questions, and we need answers if we are to reduce risks. There are other problems too, like how do you get a witness to an accident to tell you what really happened? We need a framework of understanding to show us what is going on, and to lead us to techniques where we can effect real change. There are indeed many frameworks of understanding, of varying complexity, which claim to explain human behaviour and which offer techniques for change. The one we are going to use is called Transactional Analysis.

There are many reasons for this, but the main one is that Transactional Analysis (from now on, TA) is clear and easy to understand. This has worked against it in some ways, I think, because many have assumed that because it is simple it must be simplistic. Academics often prefer models that have lots of complicated scientific language, but this is what Eric Berne was fighting against when he started developing TA. My experience is that TA explains a great deal of human behaviour in a way that anyone can understand and, most importantly, it explains how behaviour can be changed. When we start talking about changing behaviour, we might be aiming at manipulation, "spin" or downright lies. But we are not politicians, and it is important here to say that neither TA nor my presentation of it in the context of the study and prevention of accidents is about manipulation. TA, like any psychological theory, can be misused, but I am not advocating that.

2.1 Introducing Fred

So let's begin with a real example. Here is a picture of a worker sorting out cabling in a roof void under construction. I have no idea who this chap is, so for convenience we will call him Fred. You will notice that one foot is on a step ladder and the other on the top of a temporary partition which is probably not meant to be taking much weight. Most of you will know that step ladders are not designed to take pressure from the side, which is exactly what Fred is doing by working off it sideways. It is not clear what Fred is using to support himself, but whatever it is it probably wasn't designed to have people hanging onto it and it may not be strong enough. The observant will also notice that although Fred is wearing a high visibility jacket, he is not wearing head protection despite the obvious risk of a fall and the presence of all sorts of projections in the work area. He is wearing soft shoes, which might be an advantage if he spends his time clambering around things but they aren't going to protect his feet if something falls on them. The very observant will have noticed that the lights are on, so at least some of the electrical cables are live. And, to add to the sense of excitement here, the extremely observant will have noticed a danger sign near one of the ladder's legs. The yellow triangle says "Danger: Site entrance". Yes, Fred has his step ladder parked across the entrance to the work area.

Let's make some assumptions. Fred is not a juvenile, and he is doing a fairly responsible job, so we can reasonably assume that he has sufficient intelligence, training and experience to identify the hazards here, or at least some of them. He probably knows that it is dangerous, and probably against company rules, to clamber around using temporary partitions as footholds. We can also assume that the company for whom Fred works has done risk assessments, and even some method statements or safe systems of work. I bet that somewhere in a file at head office, or even perhaps at the back of Fred's van, there is a document on "How to use step ladders safely."[26] So why is he working in a way which he knows is not safe? Let's learn a bit of TA to try to understand this mystery.

2.2 Introducing ego-states

1. The Adult ego-state

According to TA, each of us is made up of three different bundles of memories, thoughts and behaviours. These ego-states are called Parent, Adult and Child and so that we don't confuse these with people who are parents, adults or children, the ego-state is always written with a capital letter. A grown up person is an adult and they have an Adult ego-state but they are not the same thing. We will start with this Adult ego-state because this is probably the easiest to understand. When my Adult is working, I am in the present, the "here and now" and dealing, as best I can, with real situations and problems in order to solve them. The Adult ego-state has "behaviours, thoughts and feelings which are direct responses to the here-and-now".[27] Undertaking a risk assessment is a good example of an Adult process. I look for hazards; identify which are likely to be serious; and ask how likely it is that the hazard is going to cause harm and, if that happens, how serious that harm is going to be. Talking about the Adult ego-state is not the same as talking about an adult person. A four year old child has a functioning Adult ego-state and can address real situations in the here and now even though they are not yet an adult. In extreme cases, an adult may act without a functioning Adult ego-state. So here is an important feature of ego-states; every normally functioning human being begins from an early age to develop three ego-states and then has them, irrespective of gender, culture, time or geographical area. A child has a Parent ego-state and a parent has a Child ego-state.[28] To summarise; we all have an Adult ego-state and we are using this when we are problem-solving in the here and now.

2. The Child ego-state

The Child ego-state contains "behaviours, thoughts and feelings replayed from childhood". This applies to you, to me, and to Fred. We all have behaviours which we learned in childhood and which we continue to replay in our lives, even though we may have left childhood a long time ago. Suppose you had a father who was often angry and sometimes violent; you might learn very quickly to avoid that anger by giving in, or being nice, or not expressing your own anger. After all, when you are a child your father is much bigger and stronger than you are. In later life you might still behave like that, especially towards males who seem to be angry. Instead of standing up to them, or expressing your own feelings, you might be extra nice and pleasant.[29]

Here's a second example: I used to work with someone who responded to problems and upsets by sulking. She would withdraw, say very little, become un-cooperative, and make it clear in many little ways that she was annoyed and upset. What she wouldn't do is say why she was upset. As you can imagine, working with this woman was sometimes very difficult. I assume that as a child she learned that a safe and effective way of getting what she wanted was to sulk and pout and be difficult. I imagine that when she did this her parent-figures (who could have been actual parents, or teachers, or important carers) would become concerned and pay her lots of attention and ask her what was wrong and put it right for her. So sulking was behaviour she learned as a little girl, and as a grown woman she was still using the technique from her Child ego-state. She had a belief, backed up by experience, that this was a good way to get what she wanted. It didn't always work, of course, but it worked often enough to be powerfully reinforced. She could have approached the problem from Adult instead of Child. Suppose she was upset because one of her staff kept on arriving late for work. Instead of sulking for the next two days while people wondered what the problem was, she could have said, "I don't like it when you arrive late for work and I would really appreciate it if from now on you came in at the agreed time." At least now everyone knows what the problem is and there is a good chance that the problem can be sorted out.[30]

A lot of humour and fun behaviour is powered from our Child ego-state. In the workplace, people often joke and fool about. Within moderation that is no bad thing – but there are occasions, especially where a group of males get together, when the joke gets out of hand and ends up in dangerous behaviour. This is, known, somewhat strangely, as "horseplay."[31] In one real example, a group of workers were stabbing at each other with high-pressure air lines. It was meant as a "bit of fun", but a worker died as a result. There must have been a lot of Child ego-state energy around when this was happening. In Adult, the workers certainly knew that what they were doing was dangerous – but that is not where the behaviour was coming from.

3. The Parent ego-state

In my Child ego-state I have behaviours, thoughts and feelings which I have stored up from my own childhood. In contrast, the Parent ego-state has "behaviours, thoughts and feelings copied from parents or parent figures". Consider again the child who has a parent or close relative or teacher who is often frightening and angry. That child might decide that the only way to survive is to do what they are told and be nice to everyone. Unfortunately, since this is probably exactly what the angry parent figure wants, the child's behaviour confirms to the parent figure that getting angry and violent works, so they do more of it. As a grown up, that former child may very well give in to any anger and adopt a conciliatory attitude. But when they themselves become angry, they may well behave exactly as their angry parent figure used to behave, because that behaviour is remembered and stored in their Parent ego-state.

I remember another example from a course on TA. The lecturer told us that for many years she kept several store cupboards full of tins and packets of food that she never used, and that vaguely, somewhere in the back of her mind, was the thought that "in an emergency" these would be useful. Sometimes friends used to remark on the large amount of stored food she had in the house. Eventually someone said something that "made her" think about it. In Adult she realised how unnecessary these hundreds of tins, packets and bottles were. She had money, and there were many shops nearby where vast amounts of food were available. So why was she doing this? Where had this compulsion to store food come from? She realised that it was something she held in her Parent ego-state, and which she had taken into her own way of thinking from her parents. They had lived in a country and at a time when food was often not readily available, and where a cupboard full of tinned food might be, quite literally, a life-saver. Her behaviour was copied from a real person, a parent, but this behaviour was inappropriate for her situation. Once she realised in Adult how unnecessary this behaviour was, she cleared out the cupboards and used them for something else. In a recent interview, the German Chancellor, Angela Merkel, told a similar story. She grew up in East Germany, where food and other supplies were often limited. Now, although she earns £257,000 a year and lives in a country where supermarkets are full to bursting, she still buys more than she needs: "I still buy something as soon as I see it, even when I don't really need it. It's a deep-seated habit stemming from the fact that in an economy where things were scarce you just used to get what you could when you could."[32] Even though she knows (Adult) that she doesn't need to stockpile food, she still does it because there is a powerful message that she "has to" do it (Parent).

4. Using ego-state theory

So now we have a simple and powerful way to analyse people's behaviour in the workplace. If you are working with someone and they are dealing with problems and tasks in the here and now, in a way that looks like it is designed to get results, then they are acting from the Adult ego-state. Are they sulking, not co-operating and disobeying the rules in a sly sort of way? Are they fooling around? These are probably behaviours which they learned as a child, and they are probably acting from the Child ego-state. Are they complaining endlessly about everything and everyone, without ever seeking ways to make things better or to cope with life's annoyances? Are they stiff and formal and always bossing people around? Then they are probably acting from Parent ego-state.

Of course, it is not just unproductive or unpleasant behaviour that comes from the ego-states. If your uncle Ahmed was a kind and thoughtful man, then some of the beliefs, standards and behaviours you learned from him help to make you a kind and thoughtful person too. When, as a child, you laughed and had fun and enjoyed all sorts of things, those memories are still part of your Child ego-state too. Note too that a particular emotion does not define an ego-state. I can be angry in Adult, or in Child, or in Parent. The key to understanding that someone is not acting from Adult is that they are not solving problems in the here and now. If I am angry in Adult about something that really is happening, then I will use my anger to change things. If I am angry in Parent or Child I shall not be changing anything, just shouting or whingeing.

2.3 More on ego-states, and understanding Fred

So how does this help us to understand Fred's behaviour? One thing is pretty clear: Fred is not using his Adult ego-state to access his training and experience so that he can deal with this day's work in a safe way. If he were doing that, we would see him putting the step ladder the right way round and making sure it is not in front of a door that someone might open. And he wouldn't be working at height balancing on a temporary screen. And so on. So, if he is not using his Adult ego-state to direct his behaviour, he must be using Parent or Child. It looks as though at the level of detailed work he is using his Adult ego-state to do all sorts of things as he works in the here and now, but Fred is not using his Adult ego-state to address the larger question of how to do this work and whether it is safe to do it that way.

In order to go further we need to distinguish between two aspects of the Parent and the Child. Let's start with the Parent ego-state. My parents were concerned for my safety and welfare. Sometimes it didn't seem like it to me, but when they shouted at me and told me off at least part of that was because they wanted to teach me something so that I would change my behaviour and do things better (which meant for them usually, and also sometimes for me). I have taken this information into my Parent ego-state as Controlling Parent. Behaviours, thoughts and feelings from my parents' behaviour has become a Controlling Parent ego-state in my mind. You can see how this is very useful; I don't have to work out every rule for myself, because my parents have already told me what to do. *"Be careful crossing the road!"*; *"Don't touch the plates when they are hot!"* *"Be careful with sharp knives!"* and so on. So when I approach a busy road to cross it, I don't have to work it out from scratch – my

Parent ego-state tells me that it's dangerous and what I have to do is look right, look left, look right again. These beliefs and behaviours are a useful part of my Controlling Parent ego-state. But there are some not so pleasant and useful bits too: if my parents told me I was a stupid boy, then that may well become a part of my Controlling Parent ego-state too and I can be pretty horrible to myself with it. Have you ever dropped something, or forgotten something, then cursed yourself and said something like *"Oh, you idiot!"*? That was your Controlling Parent ego-state telling you off internally just as one of your parents did. It is not Adult behaviour because it doesn't achieve anything. Calling yourself names and feeling bad about yourself does not help you cope with the problems you face. You just feel bad.[33]

Perhaps this is what Fred is doing. Perhaps he is feeling angry and annoyed, and he is busy telling himself off. He might be saying *"Oh for goodness sake! Don't bother with all this stupid Health and Safety stuff! Don't be such a baby – just get on and get the job done!"* Note that "all" in Fred's thinking. Parent ego-state messages often come across as universal or grandiose statements, and Fred might be saying to himself something like *"Real men don't worry about all that nonsense!"* or *"None of this safety stuff is important!"*. Neither of these statements is true, of course, but if they are part of Fred's Parent ego-state they will be important to him and they will seem true to him.

Or perhaps Fred is acting from his Child ego-state. In TA we distinguish between the Free Child ego-state and the Adapted Child ego-state. The first is when we do what we genuinely want to do, without being "adapted" to anyone else's needs, wishes, wants or demands. The Adapted Child ego-state is the opposite of that, when we react to what others want us to do, or tell us to do. A child can adapt positively or negatively. As a child and in order to survive, I may give in to those who keep ordering me around, and so I am obedient and compliant. I am positively adapted. Or I may rebel against all these bossy people and refuse to give in and do what they want. This is behaviour from the negative Adapted Child ego-state. Sometimes children alternate between these feelings and behaviours, so they may initially do what they are told, and then get their own back later. And this is behaviour which we all carry with us as adults. So Fred may be acting from Adapted Child, either positively or negatively. If he is in positive Adapted Child, then he's adapting to an established way of doing things. We can't see whether there are other people working with Fred, but presumably there are, at least some of the time. Suppose that this company and this working team has a very poor safety culture, in which the attitude is *"Anything goes here, mate – we don't bother about all those sissy rules about stepladders or hard hats!"* Suppose there is a supervisor whose message is *"Get the job done, any way you can. Don't come to me with problems, I want solutions. And don't even think of taking that H&S stuff seriously – just get this job done on time and we'll get our bonuses."* These are like our own Parent ego-state messages, but they come from other people's Parent ego-states. If this is what is going on around Fred, then he may be in positive Adapted Child, doing what all the others want, even though in Adult he knows if he thinks about it for a moment that what he is doing is not safe and the way he is doing it is not safe. This is what we mean by "peer pressure" among workers. Many employees, and Fred might be among them, will quickly go into positive Adapted Child ego-state in a group and go along with what the others say and do.

In Chapter 8 we shall be looking at safety culture, how to improve it, and how to achieve behaviour change in the workplace. We will be able to understand that much more easily now, by understanding why people do not always behave rationally, logically, or even sensibly. We don't spend our whole working lives living in our Adult ego-states. Our Parent and Child ego-states are very important and we cannot exclude their energies for very long.

Anyone who has worked with students will know that individually they are usually charming, thoughtful and sensible. But get them together on a night out, and almost anything may happen. One or more emerge as leaders (so they are almost certainly using Parent ego-state energy) and all the others go into Adapted Child. There are very powerful forces at work here that push us towards agreeing with the accepted views of the group, and it can seem very frightening if we stand out against the others.[34] Each individual's Adult ego-state is swamped by the "team spirit" and people do things which they would otherwise never normally do. Alcohol or drugs help to stifle first the Parent voice which might otherwise be uttering dire warnings about some proposed behaviour, and then the Adult, which can see clearly that this is not actually such a great idea. The resulting behaviour is at best rowdy and noisy, and at worst dangerous, cruel and damaging. And afterwards the individuals, as individuals, say "I don't know what came over me ..." Courts are full of defendants whose excuse for some gross behaviour is that they've never done it before and don't know why they did it then. If they did do something silly or worse when they were drunk, then it may have been because their Parent and Adult ego-states had been intoxicated into submission leaving the Child to do whatever he or she wanted to do without thoughts of the consequences. Females are not immune, but among a male group bravado tends to make things worse (and bravado is of course a set of Parent messages about what it means to be a "real man"). This behaviour can be seen readily among students, soldiers, stag parties and, of course, workers.

Looking at that photograph of Fred, we now have at least a possible answer to the question as to why a sensible, intelligent and trained worker might behave in a way which is not sensible, which does not reflect intelligent thinking, and which defies all the training he has ever had. He is acting, thinking and feeling, not in his Adult ego-state, but in Child or Parent, or perhaps both. He might be going along with the poor group safety culture (Adapted Child) and at the same replaying messages about being tough which he learned from his parents (Controlling Parent). We will be considering later the question of how to achieve change in behaviour, but we should already be able to see that, in a case like that of Fred there is not much point in getting angry and shouting and threatening. If we do this, we set ourselves up as an angry controlling parent figure, and our transactions will come from our Controlling Parent ego-state. This will invite Fred to react back to us from his Adapted Child ego-state, for that is what he probably did when he was a child faced with an actual parent. But we want Fred to be working from Adult, not from Adapted Child. How do we achieve that? Let's look at some more TA theory.

2.4 Transactions, discounts, frame of reference, contracts

2.4.1 Transactions
Eric Berne, the founder of Transactional Analysis, defined a transaction as the unit of social intercourse[35] and the formal definition of a transaction is "a transactional stimulus plus a

Chapter: 2 Psychological Explanations

transactional response."[36] What this means is that you do or say something with me in mind and I respond. A transaction can be words, a smile, a touch, a glance, a sigh. Although most transactions are verbal Berne knew and pointed out that the non-verbal communications that go along with what we say are a crucial part of the meaning. The tone of voice in which something is said, the intonation and volume, provide more meaning than the words themselves. To see the truth of this, imagine all the ways in which someone might say "Oh, it's you". This could be said with furious anger by someone who has just discovered that you have been parking your car in their reserved parking space all week; or it could be said with delight by a friend who opens their door to find you there. TA defines many different kinds of transactions, and very interesting and useful it is too – but for the moment we will restrict ourselves to understanding how transactions relate to ego-states.

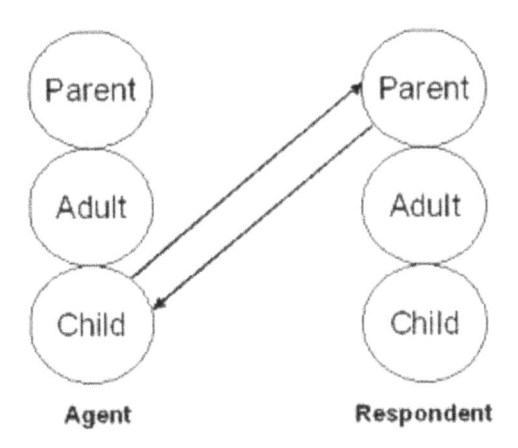

```
   ( Parent )         ( Parent )
                ↗
   ( Adult  )        ( Adult  )
           ↘
   ( Child  )         ( Child  )

     Agent             Respondent
```

In this diagram each of the two stacked circles represents a person. On the left, the person starts with a transaction from their Child ego-state. This is represented by the upper arrow. The transaction is aimed at the other person's Parent ego-state, so that is where the arrow points. The person on the right then responds, from Parent to Child. We don't know what the transaction is, but if it is verbal, an example might be

1. "Where are my shoes?"
2. "How on earth would I know?"

This little transaction could come from other ego-states, and you could have fun imagining what they would sound like and how they would be drawn. How different would it be if the initial transaction came from A to A and the response was from C to P? Remember, we cannot tell from the written words where the transaction comes from or which ego-state responds. We need to see the speaker's behaviour and hear the tone of voice. Even phrases which might seem linked to a particular ego-state are not infallible guides. A supervisor might say "Will you do the early morning shift tomorrow?" and we might suppose that this could come only from Adult. But try saying it in a sarcastic tone of voice with the emphasis on "you" and with teeth and fists clenched, and you will see that Berne was right when he saw that the words that we say are only a small part of the message. This insight is enormously useful when we interview witnesses as part of the accident investigation process, or when we aim to change behaviour in the workplace.

In the workplace, the three main transactions we are interested in are Parent - Child; Child - Parent; and Adult - Adult. Many managers, supervisors, foremen and officials of all kinds will tend to transact from Parent. After all they are the parent figures, the ones in charge, the ones who know best, etc. Some, the better ones among them, will transact, as far as possible, from Adult. Transactions starting from Child might seem to be a fairly remote option in the workplace, but as we have noted, there could be a surprisingly large number of transactions initiated from Child, in the form of jokes, horse-play, teasing, and so on. In fact it is very effective to say something that might otherwise sound like a Parent criticism in the form of a jokey comment from Child. But transactions starting in Child are mostly between people of the same status, so workers will tend to joke with each other, and managers with other managers.

You might like to notice two things here. Firstly, we cannot say in advance that a transaction from any one of the three ego-states is wrong. An initial transaction might be totally inappropriate in the circumstances, and the response could be inappropriate too, but none of the ego-states is in themselves bad or wrong. If I see a colleague walking out from behind a stack of boxes and I can see that there is a fork-lift truck speeding round the corner towards him, I am going to automatically engage Controlling Parent and shout: "Watch out! Stop!" This is not the time for Adult to Adult transactions. I want the other person to go straight into Adapted Child and do what they are told without question. (This, by the way, is why sergeants in the army spend a lot of time shouting at the soldiers. The men are being conditioned to respond in positive Adapted Child ego-state to orders from the officer's Parent ego-state. The point is that in the heat of battle, when things are going wrong and people are stressed and frightened, they will still obey orders even if their Adult ego-state is telling them that running across that piece of ground straight towards the enemy is definitely not a good idea.) But in most circumstances, shouting Parent commands at my colleagues is not going to get me very far.

Secondly, as in this example, when we initiate a transaction from our Parent ego-state and intend it for the other person's Child ego-state, this is very powerful and we will almost always "hook" the ego-state we aim at. It works the other way too – if we come from Child we are likely to hook a Parent. But we have choices. We don't have to respond from the ego-state which the other person is trying to hook. As an example, I remember one occasion when I drove my car into a public car-park and stopped in an empty space. As I got out of the car I heard an angry door-slamming noise and a red-faced man came towards me. "What on earth do you think you are doing?" he shouted (or something rather ruder but along those lines). "Any idiot can see that I was trying to reverse into that space!" Fairly obviously he was transacting from Controlling Parent, and I didn't like it. I could have transacted back from Adapted Child, apologised and grovelled, and offered to move my car. I could have reacted from Controlling Parent myself and we would probably have ended up banging each other on the nose. But I chose, quite deliberately, to transact from Adult. So I said, quite calmly, "I did not see that you wanted to reverse into this space". He was still angry for a few more transactions, but I stayed in Adult and this gave him the opportunity to calm down and, from Adult, find a solution to the problem. So, in the workplace, if you are faced with someone who is angry, you can defuse that situation by making sure that you transact from Adult intending to hook the other person's Adult. The theory says that they will either give up and walk away, or come into Adult so that they can transact with you. The same technique works with someone who is sulking and withdrawing into Child. Transact to them from Adult and they will either come out of Child or give up the contact.[37]

There is something else in human behaviour that is explained very neatly by transaction theory, and it is called the ulterior transaction. Sometimes someone says something to us and we do not hear what they *say*; we hear what they *mean*. This again supports Berne's insight that communication is not just words, but gestures, movements, etc. So the words may "say" one thing while the behaviour says another. Here again is a real example: one summer day I was walking along beside a small road on a university campus. A smart sports car drew up beside me, and a lady wound down the car window. She leaned towards me and said "Good morning. I don't suppose you know where the physics department is?" Now I knew what she meant, and she knew that I knew what she meant, and so I politely directed her to the appropriate building. And as I continued my walk, I reflected that if I had taken the meaning of what she *said* rather than what I assumed she *meant* I might have found it rather strange that she had driven to the university in order to inform me, a complete stranger, that, in her opinion, I did not know where the physics department was. Of course the question mark in her voice told me that what she meant was "I don't know where the physics department is. If you know, please tell me." Now there is nothing wrong with this kind of polite verbal shorthand, and when someone points at the empty seat next to me and asks me if anyone is sitting there I'm not going to be daft and recommend that they get some spectacles because it's quite obvious that there is no-one sitting there.

However, sometimes ulterior transactions can be less benign. There are two possible problems. I might say something which, in terms of words sounds as though it comes from one ego-state when my tone and behaviour clearly shows that my transaction comes from another. In this case I intend the ulterior transaction to be the one the person hears and responds to. Or I might really intend my transaction to come from an ego-state, with no ulterior intended, but the other person assumes that the real meaning is something different. I might for example say something to someone from my Adult ego-state, but they might hear it as though it came from my Parent, which could give it a completely different meaning. Let's look at some examples to make this clear.

Imagine that your boss sees you come into work rather late, again. She might say "Oh, I'm glad you turned up today" in a sarcastic and unpleasantly mocking tone of voice. What she says in words is one thing; what she means is quite another. What she probably means, and what you probably understand is *"Oh, you're late again, I see?"* And because she uses an ulterior you know that she is probably rather annoyed with you. She transacts from Controlling Parent and intends to pull you into Child where you will probably feel uncomfortable. Now let's look at an example of someone hearing an ulterior transaction that is not intended. Suppose someone I want to see comes into the office. I might say "Oh, hello, I've been waiting to see you." And let us suppose that this is said from Adult and intended from Adult to Adult. I mean exactly what I say – I have been waiting to see you (and now I'm glad that you are here). But the person to whom I say this might understand me to mean *"You are late!"*. This is not what I said, and it is not what I meant, but that doesn't stop the person I address from hearing it as an ulterior. The concept of the ulterior transaction is one of the most important ideas in communication between people. The wise manager will be aware of this when he transacts with workers (and indeed with other managers). And anyone conducting accident investigation interviews can improve their technique by being aware, not just of what they say, but of what the other person might hear them say.

I hope that this section has added to our understanding of human behaviour. With the material we have learned about error theory, we now have some powerful models to understand why people behave as they so often do. We shall be using all this theory again when we come to consider what to do about changing behaviour in order to prevent accidents.

2.4.2 Discounts

Imagine a driver in his car one day. He notices, or thinks he might notice, a burning rubbery smell, but he dismisses it and ignores it (*I'm just imagining it*). He carries on driving and thinks about something else, perhaps a film he watched last night. After a while however, the smell is still around, and in fact it's worse, so he grudgingly acknowledges that there actually is a burning smell. But, he tells himself, it is of no significance. It's probably that old car in front, or some idiot in the field over there burning some tyres. It's nothing to worry about. Well, a few miles further on he can definitely smell burning, and when he stops at a road junction there are wisps of smoke coming out from under the car bonnet. Now there is certainly a problem, but he tells himself that there is nothing that can be done about it. A little later, he is getting more anxious and wishes that something could be done about the smoke which is now pouring out of the engine; perhaps the garage might be able to sort it out, but he can't. So he continues to do nothing, and eventually the engine bursts into flames. Now he really has a problem.

Imagine a work situation. A construction worker is told to go and fix some temporary roofing sheets on a house being built. There is no scaffold, so he decides to use a ladder. It is raining very hard and the ground is muddy and slippery, but he goes and fetches a ladder anyway. The only one he can find has been made of bits of spare wood nailed and tied together, and it is quite old. Also, it is not long enough for the job he wants to do. The worker ignores all these problems, and starts to climb the ladder. One of the rungs is wobbly and loose, but internally he tells himself *"Just carry on, it will be alright"*. Then his foot slips because the wood is wet and his feet are muddy, and he knows he has a problem; but he carries on anyway because there is nothing that can be done about it. *"This is just the way things are on this construction site."* And even if he accepted that perhaps it doesn't have to be like that, there is nothing *he* can do about it. If he tries to find a better ladder he won't get the job done quickly and the foreman will shout at him and he might not be hired again. So he carries on. And eventually, of course, the ladder slips or gives way and he falls to the ground. Let's hope that he isn't too badly hurt.

In both these examples, a potentially serious accident could easily have been avoided. The car driver could have been alert to the burning smell, pulled off the road, and either done something himself or phoned for help. Inconvenient, yes, but better than blowing up the engine. The worker could have found a better ladder and had someone else working with him, or waited until the weather was better, or spoken to the supervisor and told him that this job was not safe at the moment and with this equipment. But there were powerful reasons why he didn't want to do that, so he ignored the problems, and ended up injured or dead. What both these people were doing was discounting reality. They started by ignoring the evidence all together (*there isn't really a burning smell: nothing wrong with using this ladder in the rain*). At some stage the evidence became too strong so, yes, alright, there is a burning smell and the ladder is a bit dodgy *but it is not significant*. This is the next level of discount; we accept that something is going on, but ignore its importance. When the problem becomes worse the driver and worker each know that they are in a dangerous situation. They have [1] accepted there is something going on (*something is wrong with this engine: the ladder is not fit for the job*) and [2] it is significant (*the engine could overheat and be damaged: the ladder is likely to break or I could slip*) but they refuse to accept [3] that something could be done about it (*these old cars always run hot: that's the way we work on this site*). Finally, they accept that something could be done, but [4] discount their ability to do anything themselves (*I can't do anything about this problem*). Here it is in summary[38]

Level	Type of discount	Typical thinking
1	Existence of problem	*Problem? What problem?*
2	Significance of problem	*Happens all the time, nothing to worry about*
3	Possibility of action	*Nothing can be done about it*
4	Personal ability to act	*I can't do anything about it*

This is the anatomy of an accident; in the first example the loss was financial, in the second, injury. Why did the driver and the worker carry on? For that matter, why did Fred work in what he must have known in Adult was a seriously hazardous manner? We have sought one explanation in terms of ego-states, and now we can combine that with discount theory. In all these cases the Parent and / or the Child ego-state is taking over and the Adult is being kept quiet. Therefore the person is acting, not in the here and now but according to ideas, behaviours and actions learned from the past. And therefore, important information happening in the here and now is being ignored. That is a discount. When we discount reality at Level 1 we don't even see that there is a problem; at Level 4 we know that there is a problem, that it could be serious, that other people might be able to solve it, but I can't. Of course, if we discount at Level 1, we are discounting at levels 2 and 3 and 4 as well. If I am not even acknowledging the existence of the problem I can't even begin to engage in solving it. If I am discounting at level 2 then I am also discounting at levels 3 and 4, and so on.

Let's relate discounts to ego-states:

- I may be seeing reality, not in the here and now, but from my Parent or Child ego-states. This means that I am ignoring parts of what is going on around me, like smoke coming from the car engine.
- I might be blanking out my Child ego-state. If I am not listening to my Child I might ignore my own needs and wants, as in the example of the worker and the accident with the ladder. He wasn't thinking of what he needed or what was safe for him; he was concentrated on doing what others wanted him to do.
- I might be shutting out my Parent ego-state, and with it all the useful rules and beliefs that I gained as a child. So I discount my own ability to do things to solve a problem.

So why do we do this? One answer is that ego-states are a part of our being. We act all the time from one or more ego-states, and sometimes we act from something in an ego-state that is negative. Stewart and Joines have this to say:

> "Whenever I am coming from any negative ego-state part, I am discounting. And whenever I am discounting, I am coming from a negative ego-state part. The one idea defines the other."[39]

This is what Fred is doing. He is discounting the evidence of the here and now and working away ignoring the very considerable hazards in the environment – many of which he has created himself. We speculated earlier about what he might be replaying to himself from Parent or Child, and without actually seeing him or talking to him, we can't say any more. But we do know that he is not solving problems using his Adult ego-state. Problems? What problems?

In the workplace, discounts are a major cause of accidents and unsafe behaviours. The typical worker or manager will spend a lot of time discounting existence, significance, possibility of action or possibility of personal action, and this will be accompanied by the sorts of Child or Parent ego-state messages that we have already noticed, such as *"All that Health and Safety rubbish is a waste of time"*. Or the internal message can be quite specific: *"Oh yes, that stack of pipes is perfectly safe"* when an assessment from the Adult ego-state would show that it was really quite unsafe. Unsafe behaviours are always accompanied by a discount, along the lines of *"Oh, it will be alright ...."* One of the reasons why H&S training works is because it invites people to look at behaviours and conditions with a new eye. Behaviour Based Safety is particularly effective here, because observers use a check-list to look at the behaviour of other workers, which should reveal any discounts. They can then be challenged.

2.4.3 Frame of Reference
We make sense of life by seeing it through what TA calls a "frame of reference".[40] This is a framework of assumptions, beliefs and behaviours which we use to define ourselves, other people, and the world we live in. Each of our ego-states is involved in this process. When something unusual or distressing happens, we attempt to understand it in terms of our frame of reference. That may be successful, in the sense that our frame of reference allows us to interpret what has happened, make sense of it, and frame an appropriate response. For many

people, TA is itself a frame of reference, or at least a large part of it because it helps us understand behaviour. So, for example, if someone gets angry with me, I can understand this in terms of ego-states, transactions, discounts and several other TA models such as anger stamps and passive behaviour.

A witness in an accident investigation has a frame of reference which will help to determine how they understand what happened to them. We may have very little idea what their frame of reference is, and that makes it difficult sometimes to understand what they are telling us. This is why, in interviewing witnesses, we are wise if we stay in Adult and ask for Adult information. If the witness tells us that something happened, and they describe it, that is something we have to accept as the best evidence we have. We then check it against other evidence, such as the statements from other witnesses.

Let me give you an example. For many years I used to buy petrol at a small Shell filling station which happened to be on a route I used frequently on my way to the university where I worked. It was small and not usually busy, and the staff were particularly friendly. Being a Shell station, a lot of things were bright yellow, and there was a large sign with a shell on it. One day I drew in for petrol as usual. I filled up my car, and went into the shop to pay. I presented my credit card, and also my Shell points reward card. The lady who served me was someone I had seen there many times, and she recognised me too. Everything went along as normal until I presented my Shell reward card, when she said "I'm sorry, but we aren't a Shell garage any more." Suddenly I saw what I had up to then completely missed; the garage was now red and the word Texaco was plastered all over it. When she gave me this information, I saw at once that this was no longer a Shell garage and it was now red instead of yellow. But before she told me that I had spent several minutes there without seeing the change. My frame of reference told me that this was a Shell garage. I wasn't expecting it to suddenly change, so I didn't see what my mind didn't expect to see. Now just imagine that I had not presented my Shell rewards card; I would have paid for the petrol, said goodbye, and driven off. If, for some reason, I had been required to give an account of the events that morning I would have been absolutely certain that I had bought my petrol at a Shell garage. I would have sworn it on oath in good conscience.

Later on we shall consider the importance of an organisation's safety culture. An organisation can very quickly swallow us up, so that we become used to the ways that things are done. Fred seems to be working in an organisation where almost anything is allowed, however hazardous the operation. Once we become part of the group thinking and behaviour, it is very difficult to change either our frame of reference or the organisation's way of doing things.

2.4.4 Contracts

Another key concept in TA is that of the contract. The idea is not unique to TA of course, but the emphasis probably is. Any psychotherapist working with the TA model will agree a contract with their patient before undertaking any work at all. Otherwise, who knows what they are aiming to achieve or how long it is likely to take; and how will they know when they have achieved it? But contracting is not just for psychotherapists. I believe that a good understanding of contract theory would revolutionise the way that management functions in any context, and that includes Health and Safety.

We all spend a lot of our time and energy trying to get other people to do what we want. In the home, children want things; parents want things; partners want things. At work, there are conflicting interests between managers, workers, HSE people, customers, and so on. Now there are many techniques for getting what we want. When people have power, the obvious way of getting things is simply to demand it ... or else. The Parent ego-state of the person with power (parent, boss, politician, industrialist, banker, mafioso) says *"Do it – and do it now and do it my way – or else"*. With enough power behind it, such a technique usually works. If you are the boss, you can simply tell other people what to do. If you are a parent, you are much bigger and stronger than your children, so you can tell them what to do (until they grow up and are bigger and stronger than you are). But there is a problem. What happens if you transact from your Parent ego-state, pull the other person into their Child ego-state, and force them to do what they don't really want to do? The answer is resentment. Children may "have to" do what their parents want, but they will resent it and, when the time comes, get their own back. Workers may toe the line if they think their wage or their job depends on it, but they are likely to be resentful, uncooperative and surly. And, again, they may get their own back in small but infinitely frustrating ways.

On the other hand, if someone does not have power, different techniques can be employed. Whining, sulking, withdrawal, crying, refusing to eat; these are some of the means by which children (who very often feel powerless) succeed in getting what they want. As adults we are tempted to still use some of these techniques that we learned as children. Workers are not likely to refuse to eat, but they may well withdraw and "sulk" in various ways if they cannot otherwise get what they want.

Eventually the other person may give in ... "Oh well, all right then ..." But again resentment is likely to build up.[41]

Or, we can ask for what we want in an Adult to Adult transaction and agree it with the other person. This leads to a contract. The problem is that not many children are used to Adult to Adult contracts. As we suggested above, they are likely to be more used to power plays and manipulation, and that is what then gets reproduced twenty or thirty years later. But children have perfectly good Adult ego-states and do agree all sorts of things with their friends on an equal footing. And it would be a wonderful thing if parents stopped using Parent all the time and agreed contracts with their children. And with everyone else, for that matter.

So, a contract is an expression of what I want, and an agreement from the other person. It is not forced or demanded. A simple example, one that I have faced many times, is the question of housekeeping, keeping the workplace tidy. This may sound deeply uninteresting, but slips, trips and falls are the major cause of accidents in the workplace. Keeping the workshop or the construction site tidy goes a long way to preventing many of those accidents. As a consultant, I often turn up at the site and find that there are bricks and bits of wood and piles of rebar and goodness knows what all over the place. It isn't safe. What do I do? Well, I can shout at them and tell them to do what they are jolly-well told, and that they must tidy up and keep it tidy. We can start a poster campaign and put notices up all over the place demanding the approved behaviour. Workers can be threatened and cajoled. Since I and the managers have authority, this will work for a while. But very quickly, the situation will revert to the usual chaos. So now I have less credibility since I achieved nothing – I lost. Next time I arrive on site it will be much more difficult to get the workers to do what I want.

Now suppose that instead of transacting from Parent and telling them what to do I invite the workers to make a contract with me. So I might have a brief meeting with them all and say that what I would like is for them to keep the site clean and tidy because they will undoubtedly be safer that way. And I can ask if there is any problem with that? Now they can tell me all about the problems they experience, and we might find that although in theory the site supervisors are all in favour of keeping the place tidy, that actually no procedures are in place, no-one takes responsibility and that the pressure on the workers is to get on with the job in hand, not to do something about the bits of dry wood piled up next to the welding bay. We can have a discussion and solutions can be proposed. Both management and workers are likely to be involved. But what I am aiming for is a contract in which the workers agree to take responsibility for keeping the site safe and tidy. "Yes, OK, we will do that." This is much more likely to achieve lasting results. Experience tells me that there will be an improvement, but that the discussions and contracting needs to be continued for effective change to take place. After a while, the workers will see it as part of their routine to undertake good housekeeping. The contracting process invites Adult to Adult transactions, which imply equality, respect and responsibility. These are precisely the sorts of changes we need if safety culture is to improve, so contracting is an important part of that process.[42] Another reason why BBS works is because it involves this kind of mutually respectful contracting.

Some final points about contracts. They must be mutually agreed. Something that is imposed by one person on another is not a contract. A contract should be in positive language – it should undertake to *do* something, not to *not do* something. And it needs to be achievable. Has anyone in the world ever done this? Can I do it? What will be the cost of achieving this contract? What needs to be done? And how will I or we know when we have done it? Comparatives like "better" must be rigorously excluded unless they refer to some measurement point. A contract which says "We will keep the place tidier" is pointless. How much tidier? Tidier than what? When? Contracts also need to be moral and legal; something that will increase production at a great cost to the environment is not acceptable. And remember it is the bits of the contract that are fuzzy and undefined where the problems will arise. For example, it is great to have a contract that says what we are going to do, how we are going to achieve it, and how we will know when we have achieved it. But if we don't agree *when* we are going to achieve it, the whole thing may be useless.

And what happens if the contract doesn't work? The first response should be to find out why and renegotiate. However, there may be situations where legal requirements take precedence. For example, there are rules in the UK about workers wearing PPE, and it is not enough for the managers to encourage and contract and then ignore the problem if it is not solved. Managers have a legal responsibility to ensure that workers are protected from hazards and that they are not put at unnecessary risk, and it is no defence for a manager to say that the worker agreed to wear the PPE but then didn't. So there may well be situations where an essential part of the contract is an understanding of what the consequences may be if it is not kept. Two people can agree on a course of action and still be aware that if the contract fails then a Parent instruction will follow. And if that fails then there may be disciplinary action.

Contracts are not a soft option for avoiding responsibility, but an effective means of achieving mutually-agreed change.

In the following chapters we shall continue to use these concepts to illuminate our understanding of accident investigation and prevention.

Chapter 3: The Advantages of Good Accident Investigation

3.1 Why should we investigate adverse events?

Why is it important to investigate accidents and incidents? There are of course many reasons why you might undertake an adverse event investigation, and there are many lists in various sources. Here is a selection, in which the first ten apply to all adverse events (accidents and incidents) while eleven to fifteen will usually apply only to accidents.

We might undertake accident investigation with the following objectives:

1. Reducing the likelihood and impact of future adverse events
2. Improving and correcting the SMS
3. Improving operating efficiency and increasing profits
4. Reducing danger to individuals, eg
    a. employees
    b. customers
    c. visitors
    d. cleaning and maintenance staff
    e. the public
5. Maintaining good morale
6. Educating staff at all levels
7. Preventing the loss of good staff
8. Identifying violations of company procedures
9. Undertaking research

10. Complying with company rules and / or external regulations on accident investigation
11. Preparing information for regulatory and / or government agencies
12. Preparing answers for
    a. the media
    b. local residents
    c. local organisations
    d. shareholders
    e. insurers
    f. workers' compensation organisations
13. Providing protection against litigation
14. Making it look better
15. Finding someone to blame

We shall now consider each of these in turn.

3.1.1 Reducing the likelihood and impact of future adverse events
It is often said that we investigate accidents in order to stop them happening again. Well, yes, of course we want to do that. The lessons that we learn from investigating adverse events can have a profound effect on the future of the workplace. But, as we learned when we looked at the theory of latent errors, we are not focused on the prevention of that particular accident happening again in the future. Surely we want to achieve more than that. We want to discover latent errors in the system and correct them so that no adverse events result from the fault.

3.1.2 Improving and correcting the SMS
Adverse events show us that something is wrong with the system, so by feeding back information from an investigation into the SMS we can improve all sorts of things and we will, we hope, make the process safer and prevent, not just that accident

happening again, but other potential accidents too. We can also make sure that our barriers and defences are still in place and working effectively, so that the negative effects of something going wrong are reduced.

3.1.3 Improving operating efficiency and increasing profits
Investigating adverse events and learning from the mistakes that have been made should definitely improve operating efficiency. There is always a temptation to get the job done more quickly and, therefore, it is assumed, more efficiently. Workers may take unsafe short cuts to get work done speedily, and in the short term they may succeed in producing more widgets an hour by removing the machine guard. Supervisors may reduce down time and lost production by sticking a bit of metal into the broken interlock guard so that the machine can still be operated. But of course, when the bit of metal slips and a worker is killed when her head is caught in the closing machine (a real example) there is much more disruption to the production process than there would have been had the guard been repaired in the first place. One reason why we investigate adverse events is to prevent operating and working practices (undesired circumstances, unsafe acts) that are, in the long run, inefficient.

The aim of most organisations is to make money; or at the very least to create enough income to keep going. The prevention of accidents is a process that takes time and effort, and in itself costs money. But accidents also disrupt the wealth-creating process, so they too cost money. As always, there is a balance between the resources spent on reducing risk, and the gain from the anticipated risk reduction. Spend too much on risk reduction, and the financial strain may bankrupt the company. Spend too little, and the company may go out of business as a result of a catastrophic accident. Many do. Here we get into the realm of "so far as is reasonably practicable" and other balancing acts. In some cases legislation lays down precise rules about what has to be done, or specifies that all that is actually possible must be done, but in most cases it is one of the responsibilities of management to decide how much is to be invested in reducing risk and by how much. Clearly, if a small investment of resources (time, money, effort, inconvenience, etc.) is going to reduce risk by a large amount, it is worth-while. And the company might also lay itself open to prosecution if it did not take simple and inexpensive measures which would greatly increase safety. At the other end of the scale, an enormous investment of resources which might result in a small risk reduction would probably not be considered sensible. One of the objectives of adverse event investigation is to gain detailed understanding of the plausible risks and to help make that judgement about what might be done to reduce risks, and at what costs. In short, information from accident investigation can and should be a powerful tool which management uses to reduce operating costs.

3.1.4 Reducing danger to individuals
Reducing danger is a clear and laudable objective, and we achieve this in the process of identifying hazards and reducing risks. When there are adverse events in our place of work this indicates that the SMS is not working properly, which means that adverse events are going to continue to happen; and that means that sooner or later someone is going to get hurt or there will be some other loss. It is natural to think immediately of employees, but do not forget that managers, secretaries, delivery drivers and all sorts of other people are employees too. Then there may be visitors or customers in the building at various times. Do not forget maintenance staff and cleaning staff, who may be contracted in from outside and who may be on the

premises at times when most other people have gone home and when normal support systems may not be operating. Lone workers and people operating at unsocial hours may be more at risk of accidents than regular staff, and the consequences of accidents may be more severe if help is not immediately available. Then there is the general public; they may know nothing of your enterprise except the outside of the buildings, but if something serious goes wrong they may be affected.

3.1.5 Maintaining good morale

If you have ever been involved in a serious accident, you will know that there is always a negative effect on morale. People are shocked, distressed, and jolted out of their normal routines. Work may stop for a while. People take time to adjust to the new reality and spend energy on attempting to explain what happened. There may be several direct negative results.

Firstly, people work more slowly (and perhaps more carefully) for a while. This phase may not last long but can be significant. In extreme cases people may feel traumatised and unable to continue working at all, so valuable workers will be lost to the enterprise.

Secondly, people are wondering, worrying and working out answers so they are not thinking and acting as they usually do. This is a situation in which another accident is likely to happen.

Thirdly, there may be a loss of trust. This is especially likely if people think that the accident could easily have been prevented – and in most cases that is so. Why then did management not prevent it? If the likely answer seems to be that management was unwilling to invest the resources to make the work safe and prevent the accident, then morale is likely to plummet.

Management may now be very concerned and showing every sign of taking the health safety and well-being of their workers seriously – but was this the case before the accident?

If management does not take the actions necessary to prevent disasters, and if as a result there is a serious accident, or a series of unpleasant incidents, the ulterior transaction message from management to workers may be, or may be heard as, *"we don't care about you"* which equates to *"you are not important"*. I do not mean that management is going to actually say this, or put it in a memo. In fact quite the opposite is likely to be the case and management will say all the right things about "Safety is our Number One Priority" etc. But we've heard it all before: the question is, do they mean it? And the only way we know if they mean it is if they do what really hurts; and that is, pay for it with money and resources. If I come to believe that, whatever they say, they don't mean it, then my Child ego state may be afraid. If "they" didn't care enough to prevent this accident, then they probably don't care enough to prevent the next one, and the next time I might be the person who is injured. All sorts of resentments build up in such a situation, often along the lines of *"If they don't care about me, why should I care about them?"* When the memo comes round instructing workers to try harder, it is likely to be met with anger rather than co-operation. (The memo is in any case often a Parent to Child communication).

Good industrial relations and good morale are certainly going to be much better if contracting between management and workers is a daily reality. A trusting relationship will have been developed in which it is not a "them" and "us" (which is to say, Parent and Child) relationship but a mutually supportive Adult to Adult relationship. In the unhappy event of a serious accident, this trust will be very important.

3.1.6 Educating staff at all levels

Another reason for undertaking adverse event investigation is to educate staff. Such learning happens passively and actively. Passively, those not involved in the investigation process will learn something important, which is that the company management cares enough about HSE to spend time and money on investigating adverse events, and not just the serious accidents (where they probably have little choice) but also incidents. This perception is a major factor in improving the organisation's safety culture. And actively, those who take part in the investigations are learning on the job and improving their skills.

3.1.7 Preventing the loss of good staff

This goes along with education and maintaining morale. Organisations where the ulterior transactions are *"we don't care about your safety or health"* will not retain the best staff. Those who can, will leave. One of the causes of the Bhopal disaster was that as safety at the plant was run down many of the better workers left. They could see that this was not a safe place to work, morale had plummeted and there were no promotion prospects. According to Kurzman

> "cuts...meant less stringent quality control and thus looser safety rules. A pipe leaked? Don't replace it, employees said they were told ... MIC workers needed more training? They could do with less. Promotions were halted, seriously affecting employee morale and driving some of the most skilled ... elsewhere"[43]

3.1.8 Identifying violations of company procedures

Company procedures are put in place as part of the process of managing risk. If these procedures are regularly violated, that means two things. Firstly, they are ineffective so risk is not being appropriately managed. Secondly, there is something wrong in the system which (a) encourages and (b) allows violations. So uncovering this information through accident investigation is an extremely useful outcome.

3.1.9 Undertaking research

In some cases an accident investigation will uncover a system failure which might apply to other sections of the plant or operation, or to similar operations elsewhere. In this case, the information gained from the investigation can be published so that others can gain from the insights. This might be in an in-house newsletter, a trade journal, or an academic publication.

3.1.10 Complying with company or external rules

The well-managed company will have rules about which adverse events will be investigated and it is important for the safety culture that these rules are known and seen to be followed. There may also be industry association standards which the company has decided to follow. There may be "best practice" guidelines set up externally, by an association or by a parent company. As always, if management says "This is what we are going to do as part of the management of HSE" then, unless they do it and are seen to do it, their protestations will be seen as an attempt to get the credit without spending the money. That will contribute to a poor safety culture and that in turn is a major factor in the occurrence of adverse events.

3.1.11 Preparing information for regulatory agencies

This probably only applies once an accident has happened. There may be many different agencies which will require a report of some kind. These might include

- Disaster management agencies
- Health and Safety ministries, departments, agencies etc
- Environmental agencies
- Police services
- Fire services
- Local authorities
- Licensing authorities
- Insurance companies

3.1.12 Preparing answers for the media and other groups

Every national and local news bulletin is full of reports of disasters, crimes and accidents. Occasionally there will be a few seconds of some heart-warming and humorous story to conclude with, but it is bad news and disasters that sell newspapers and media slots, and it is the things that go wrong that excite the reporters and news-gatherers. If there is a serious accident, and especially if there is loss of life or significant damage, then the media is going to be reporting on it. And it might be you that they are reporting on. It is therefore essential that you and your company have a well-prepared plan for satisfying the media in the event of an emergency. If something goes wrong and you do not provide information, then the media will find information to fill the void, and such information might or might not be accurate. Sometimes it doesn't make any difference, and the news hounds report what they want to report anyway; but your chances will be better if you provide honest information quickly. Vincoli puts it rather dramatically:

"One thing is certain, companies that do not consider public relations an essential element of the accident investigation and loss control process will certainly suffer the detrimental effects that the uninformed media and the public can inflict following an accident event."[44]

If there has been an accident, you need to be able to say that an accident investigation team is already at work and that your organisation will issue a statement as soon as possible and act on the findings. You will want the media to report that you take these things very seriously, are aware of the immediate implications, and will respond responsibly and realistically to the accident investigation findings. The worst thing you can do is to play it all down in the hope that it will go away – the reality will come back to haunt you. So make sure that there is an investigation team starting work even as the news cameras turn up at your front door, and that there is a sensible press release, and that whoever is giving the news to the press knows what they are doing. This is an attempt at a contract: we will keep you properly informed if you will treat us with respect. It might not work (especially in the UK where the popular press seems to take the view that as long as they can sell more newspapers they will print anything) but it is worth a try.

It is extraordinary how many highly paid and experienced managers behave like children when something goes horribly wrong. "Nothing happened and I didn't do it anyway, and anyway it was his fault" seems to be the knee-jerk response. Just as it was when mummy or teacher caught us doing something very naughty. The reaction comes straight from the Child ego-state. There are dozens of recent examples, but BP will do. In April 2010 the *Deepwater Horizon* oil rig exploded and then sank, with the loss of 11 workers, injuries to 17 others and a subsequent enormous oil spill. According to BP, there was apparently no problem – the oil spilling out into the ocean was "relatively tiny" compared to the size of the ocean, and the environmental impact would be "very, very modest".[45] And it was the fault of the sub-contractors anyway. Needless to say, this kind of response leads to contempt and mistrust as the initial denials and blame-shifting fail. So instead, have a realistic response based on your accident investigation strategy. And don't start making up excuses or providing answers until you really know what happened.

The media may be only the beginning of the story. If there is an accident with significant effects, all sorts of individuals and groups are going to take an interest and start asking all sorts of questions about you, your organisation, and what you are going to do about all of it (whatever it is). Some will be genuinely concerned and serious. Others may be on to you to see what they can get out of it. But all of them need answers and information, and that response needs to come from the accident investigation.

Shareholders and investors are also going to want to be kept informed. Customers may be anxious that there will be disruption or price rises because of the accident, so you should contact them too. Ignorance leads to anxiety and other suppliers may see a good business opportunity for new sales and approach your customers with seemingly attractive offers. All this needs to be managed calmly and competently.

Those who live and work in the area where the accident has happened are important and should be kept informed of what is going on. This is a good aim in itself, but it also prevents the growth of pressure groups which may be led or taken over by agitators. Many local groups are concerned with serious issues, even if they may seem hostile to the business organisation. Dismissing them and ignoring them can lead to lots of worse trouble.

Insurance companies and workers' compensation boards will want to know what is going on. Since they may be very important to your organisation, and since large amounts of money may be involved at some stage, you would do well to keep in touch and provide information and co-operation.

3.1.13 Providing protection against litigation
This is a complex area and countries, states and jurisdictions differ. All that can be usefully said here is that your legal advisers need to be part of the investigation process. There are so many areas where the investigator needs to know the legal status of information and evidence, and that means getting expert advice.

On the one hand the company wants to get the information that it needs to achieve its stated aims of continuously improving the management of HSE. This is not adversarial, which is to say there are no claims and counter-claims, blame and defence. On the other hand, most evidence gathered during that process could be used by individuals or organisations as evidence in a legal process. That *is* adversarial, where each side will use whatever it can to prove that the other guy is wrong, incompetent and blameworthy.

The best advice for the accident investigation team is to keep it professional. Until the investigation is completed, avoid stating opinions or interpreting evidence unless it is essential to do so.

Keep the whole process clean, calm and Adult. Later, if called as a witness, the investigator should stick to the facts and, if the investigation was a good one, that will be clear to everyone concerned.

3.1.14 Making it look better
As we mentioned above, some management teams will talk the talk without walking the walk. Their interest is only in making it all *look* better. This should not be an aim of accident investigation; the aim is to *make* it better.

3.1.15 Finding someone to blame
Finally, here is another aim which has in the past been an important part of the process, especially in those companies which accepted the Heinrich concept of the unsafe act. If you have reached this point in the book, you will know that this way of doing things is now totally discredited. Accident investigation is not about finding out who did what wrong, but about examining many possible causes of an adverse event and using that information to improve the management of HSE.

3.2 The direct and indirect costs of accidents

Accident investigations cost money. Resources are needed; for example, the investigator(s) need time, which has to be paid for even if the investigation is in-house. You may need to bring in experts or consultants, and they aren't cheap; and the company needs to provide resources for the team. Reports have to be written, produced and circulated. So if the management is going to approve adverse event investigations they will be accepting the cost implications, and they may ask why it is necessary to spend this money. Here are some answers.

Incidents are a free lesson, but accidents are usually very costly. These costs can be divided up into direct and indirect.

Direct costs might include:
- Cost of accident investigation
- Lost production
    - injured worker absent from job
    - damaged machinery
    - down time and disrupted work-patterns
- Loss of trained workers who retire early or who have to change jobs
- Cost of repairs to buildings, machines and equipment
- Cost of replacing
    - trained personnel
    - labour
    - machines
    - premises, etc
- Medical expenses
- Compensation
- Increased insurance premiums
- Fines

Indirect costs might include:
- Loss of morale in workforce
- Loss of positive image
- Loss of future contracts
- Loss of good will (through delays in delivery, for example)
- Expense of defending against litigation

Then there are the costs to the injured personnel and their families:
- Physical or mental suffering

- Disability and negative reactions from others
- Loss of income
- Disruption to family routines, plans, etc
- Inability to continue hobbies and fitness routines

These are only a few examples. Bear in mind too that many costs are uninsurable, either because insurance isn't possible (eg against fines) or is too expensive.

I once worked as a consultant with an HSE manager who was in charge of a very large construction site. One day he said, rather to my surprise, that being an HSE manager is the worst job in the world. When I asked him why this was so, he said, "Well for a long time, everything goes along nicely and there are no costly adverse events. So the big boss calls me in and says, "Nick, I employ you as HSE manager, but nothing goes wrong and we have a good HSE record, so why am I paying you all this money?" Then there is an accident. So the big boss calls me in and says "Nick, I employ you as HSE manager, now this has gone wrong and we have a poor HSE record, so why am I paying you all this money?"." This is the problem of HSE. Management is being asked to invest time and money into something which, if it works, produces nothing as an end result. It is difficult to maintain enthusiasm for this process. So it is important to bear in mind all the time that actually what a good HSE management system is achieving is the avoidance of very costly, and potentially disastrous, events.

Chapter 4: Investigating Accidents

4.1 The seven essential arts of the accident investigator

To get us started, let's list the seven essential arts of the investigator and then take a look at each in turn. We shall return to some of these later in more detail, but we can begin by establishing a simple frame of reference.

The seven essential arts of the investigator are:

- Knowing which adverse events to investigate
- Knowing how far to investigate
- Knowing who should undertake the investigation
- Being prepared in advance
- Knowing who has what authority, and using it
- Having a good investigation tool kit
- Writing a good report

4.1.1 Which events do we investigate?
For convenience, let's describe four levels of adverse event, the first three of which result in loss and are therefore accidents. Level 4 comprises near misses and undesired circumstances.

Level 1	Fatality or serious injury, major damage or loss
Level 2	Some injuries, some damage, loss, and disruption
Level 3	Minor injuries, low-level damage, minor loss or disruption
Level 4	No injuries or damage, but potential to cause them

Level 1 accidents involve major damage, disruption and either serious injuries or deaths. With these accidents we are not going to have much choice whether we investigate. The press and media will be telling the world about the disaster, and quite apart from any other good reason we might have, we will need to hold a major investigation in order to protect ourselves. It is also very likely that the law and regulatory bodies will require an investigation to be held, and there may be an accident investigation imposed by the police or a regulatory body, or both. Even if there is an external enquiry, the company will still be wise to conduct an internal accident investigation as well.

Level 2 accidents are less serious, but will still involve significant levels of disruption and loss in terms of finance, materials, production, reputation, etc. Here too, investigation is essential. Something has gone seriously wrong and the organization needs to know why.

Do we investigate after a Level 3 accident? On the one hand, there will probably not be a legal requirement, and an investigation will add to the costs and disruption already caused. So it may seem like a wise decision to clear up and carry on as usual. On the other hand, carrying on as usual may lead straight into another accident, and it could be more serious next time. If the accident was potentially much more serious, then investigation is called for. If the accident (or something very like it) has happened before, or looks likely to happen again, then an investigation is sensible. So whoever is responsible for this needs to make a decision based on what the likely cost of the accident investigation is going to be and what the likely improvement in health and safety is going to be. On balance, it is best to have a system that investigates a fair number of Level 3 accidents. Having said that, it is important that the amount of effort involved is sensible. Standard forms and procedures can make the investigation quicker and easier.

Most organizations would agree that it is impossible to investigate all Level 4 incidents. The cost of this would be far too great. On the other hand, if the HSE manager knows about near-misses and undesired circumstances, then they can keep an eye open for statistical trends and clusters. If there are no other indicators, then sample a few incidents and investigate them thoroughly; they may show that the safety management system is working terrifically well, in which case, jolly good. Or they may reveal some latent errors lurking in the system. If the number of incidents is increasing, then some at least should be investigated. If incidents are being reported from one particular area, or at a certain time of day, or amongst a group of workers, then an investigation is indicated. If the organization has critical areas, either in terms of HSE or of the activity being carried out, then it is probably wise to investigate all incidents that occur there. If there is one production area on which everything else depends, it will be essential to know if anything is going wrong here, and as early as possible. And in safety-critical industries all relevant information needs to be gathered. This is simple risk assessment, of course: the worse the outcome is likely to be if an accident occurs, the more effort and expense needs to be put into the monitoring and control of risks.

To summarize: If the answer to any of the following questions is YES, the adverse event should be investigated:

- Is there a legal requirement to investigate?
- Is an outside organization going to conduct an investigation?
- Is there major public interest?

Chapter 4: Investigating Accidents

- Is the press / media involved?
- Are there likely to be legal proceedings arising out of injury / loss?
- Could it easily have been much worse?
- Does the adverse event clearly indicate failings in risk control?
- Is this adverse event likely to happen again?
- Has it happened before?
- Is there a statistical trend or a cluster?
- Has no investigation been carried out for a while?
- Is this a safety critical area / process?
- Is this a critical production / process area?

4.1.2 How far do we investigate?
The answer to this is surprisingly simple. In most cases the extent of the investigation will be linked to the severity of the adverse event. In every case, however, an investigation needs to go far enough to reveal deficiencies and latent errors. Go as far as you can as long as the use of resources is balancing the usefulness of the evidence that you find. When you reach the point where you are no longer recovering any information that is useful, stop. No investigation can cover everything, so again the skill of the investigator is important in deciding on the allocation of resources.

4.1.3 Who undertakes the accident investigation?
This question might not have a simple answer and will usually depend on the seriousness of the adverse event. With Level 1 accidents outside experts and consultants will probably be involved, although the HSE manager, if skilled in accident investigation, might be a part of the investigation team. The local HSE manager might instead undertake an in-house investigation in parallel with the formal investigation. Level 2

accidents also call for considerable expertise and an external consultant might be engaged to lead the team which, as before, might well include the local HSE manager.

Level 3 and 4 adverse events are different. Here the company will probably not want the expense and trouble of calling in outside experts; the HSE manager will have to do the job. But they may not have to, or be able to, undertake all accident investigations, especially at Level 4 where there may be a large number of incidents. On the other hand, supervisors, foremen and others with responsibility in the workplace can be trained to undertake investigation at this level. The role of the HSE manager is then to review and if necessary supplement these reports, and to report to management. Care must be taken, however, because line managers have a vested interest in the system – they are, after all, responsible for some aspects of its day to day management. This is where it is tempting to fall back on blaming the operator without going further and examining the environment and the system. In reviewing locally produced reports, the HSE manager will be well-advised to bear this in mind.

Manuele insists that accident investigations should be run by managers because they are responsible for loss control. His point is a good one and I certainly agree that if there is no managerial responsibility, accountability and involvement, we might as well all go home. But I don't think it always has to be a manager who conducts an adverse event investigation. Supervisors and foremen can and I think should be trained to do this, for two good reasons. Firstly, more Level 3 and 4 incidents will be investigated and yield up their valuable content of information. And secondly, if as many people as possible are involved in accident investigation it becomes a natural part of the operation, not something imposed by managers on the workforce. In this way, as in so many others, investigation helps to create a good safety culture.

4.1.4 Being well prepared

It is always a good idea to have terms of reference for the accident investigation and report agreed well in advance. More will be said on this in Chapter 7. The amount and kind of preparation you will need to do depends on a multitude of factors. A senior HSE Officer friend of mine works for part of the time on a remote tropical island so when investigating an accident he might have to fly to a site by helicopter. Once there, he will have very little in the way of resources except for those he has taken with him so his planning has to be meticulous. Other investigators in similar situations might need passports, visas and complex travel arrangements. At the other extreme, in a smallish manufacturing plant the HSE manager may be working in familiar territory and be able to call on all sorts of back-up. Much less planning is necessary.

If an accident happens at a site remote from your normal workplace, you can't freeze the situation so that everything is waiting for you to arrive. As soon as the accident occurs people are going to take charge and do all sorts of things. So you might find it a good investment to make sure that local site foremen or factory or site supervisors know what to do. Training them in advance so that they can react quickly and effectively in the event of an adverse event might take a lot of pressure off the HSE manager or department.

The wise investigator will assess the situation, the likely calls on his expertise, and the sensible arrangements that need to be made to avoid frustrating delays when things need to be done quickly.

4.1.5 Having authority
The investigator needs to know what authority they have. For a simple investigation in your own work place, this will probably already be clear – but if you are not sure then it is a good idea to find out. Better to do this in advance than face a situation where someone is saying that you can't do something or other and where you don't know if they are right or not. The situation becomes more complex where there are two investigations going on at the same time, as may happen if there is a serious accident and a regulatory body becomes involved.

Once you are clear about your authority, you can use it. Once an accident has happened there are usually difficult decisions to be made at an early stage, and people may be frightened, upset or in shock. Controlling Parent is useful and calm Adult is essential.

4.1.6 The accident tool kit
Anyone who is likely to be called in to investigate an accident needs to put together a selection of useful items which can be taken to the scene at short notice. We will look at this in detail in just a moment.

4.1.7 Writing a report
Investigation is no use unless some fact-based and useful results are recorded. And recording them is no use unless the information reaches the people who need it. And there is no point in people having information if no action results. So writing, presenting and following through an accident report is essential, and will be considered in Chapter 7 below.

4.2 What do you have in your investigation tool kit?

Like the Boy Scout, the accident investigator must always be prepared. Presumably the Boy Scout has to be prepared for anything, which is a pretty difficult position to be in. For the accident investigator however, things are a lot easier, for at least they know that they have to be prepared to investigate accidents, even if they don't know what the accident will be. When the phone rings or someone knocks urgently on your office door, you should be ready to go to the accident scene as soon as possible. Getting there quickly is very important but, of course, the need for speed does not override the need for safety. As always, assess the risks and take the necessary actions to reduce or control them.

When the call for action comes, it is too late to start wondering what you might need and where it is to be found. This preparation should have been done, in an unhurried and calm manner, before any accidents have yet come your way. Most investigators have a case or bag in which they keep the tools they think they will need when the call comes. The idea is that you can grab the bag and get under way with maximum speed. The bag may be provided by the employer, or you may have your own. It may be in your office, the boot of your car, or in your study at home. But in any case, what do you need in your tool kit? Here is a list of possible contents:

- Torch*
- Camera for still pictures*
- Video camera*
- Voice recorder*
- Laptop*, power lead
- Portable mouse* and mat

- Cables for connecting cameras and voice recorder to laptop
- Stopwatch
- Hazard warning tape
- Tape measure, ruler
- Squared ruler for showing size on photographs
- Magnifying glass
- Envelopes, small boxes, plastic bags, sample bottles
- Thin latex gloves
- Tags, labels, sticky labels, post-it notes
- Pencils, different coloured pens
- Squared paper, note book, clip board
- Compass
- Prepared forms
- Specialised tools such as noise meters, gas detectors, etc.

Though they probably won't be in the bag, you should also be carrying your mobile phone (and charger or spare battery) and whatever PPE is appropriate. Also you might need ID or evidence of your authority.

This is a list of suggestions, and of course it is up to you to make the final choice of contents for your grab bag and to add anything else that you might find useful. I recommend a bottle of water and a banana and a couple of chocolate bars as an emergency energy top-up!

The asterisk against the items at the top of the list is important. This is because they need batteries. And they need batteries that are fully charged, and you might need spares, and a charger or charger lead. Nothing gives you a worse start on the job than to turn up with a camera which doesn't take pictures because the battery is flat. So if you are keeping a special camera in the grab bag, you will need to remind yourself to recharge the battery at regular intervals. And if you are going to use a camera which is in everyday use, then you will need to know where it is and also make sure that it has a working charged battery. You might also want to include spare memory cards or storage devices for pictures and films.

4.3 What you do and when you do it

Here is a check list of what you might do when an accident happens, grouped into an order of importance. Within each section the actions won't necessarily take place in that order, so of course you will use your experience and common sense. As with any checklist, this one is designed to cover all the eventualities, and not all of them will apply to every situation. Please use the list to plan your response for your own situation.

4.3.1 First things first – safety and help

- Arrive safely
- Make sure that it is safe to proceed
- Call emergency services (fire, police, ambulance, emergency response teams)
- Provide immediate first aid or other attention for injured people
- If there are shocked and traumatised people who are not physically injured, move them away from the accident scene and provide shelter
- Prevent further damage to property, buildings, machines etc

Whenever I teach this course, I ask the class what is the first thing they would do when they arrive at an accident scene. Most people say that they would help the injured or phone the medics or police, and this is of course extremely important. But in my view this is not the *first* thing you should do. First of all you must make sure that it is safe to proceed. It is not a good idea to rush in to help the injured if the roof is just about to collapse. If someone is injured, make sure, as far as possible, that whatever injured them is not going to injure you. Think energy sources, falling objects, gasses and fumes, etc..

If you are investigating a Level 3 or Level 4 adverse event, you can probably do everything yourself. But at Levels 1 and 2 let's hope you are not on your own. For the more serious events you should have a practised emergency response plan in place, and there should be people with responsibility for doing some of the things that need to be done. Your job as investigator may be, at this early stage, to check that these things are being done and to react to any unforeseen problems. During this process, you will be able to gain an initial impression of the accident scene, as the priorities set out above merge into the actions set out below. If emergency services are at the scene, they must have priority, so that further injury and damage can be prevented. However, you may be able to co-operate with them to begin the investigation before they have completed their work, if it is safe to do so.

4.3.2 Once the scene is safe
Here is a list of things you will probably need to do once the site is safe:

1. Secure the area so that evidence is undisturbed. Restrict access to those who need it and keep others away
2. Inform whoever needs to be informed about what has happened
3. Make an initial tour of the site (and keep safe!) to get an overall impression of the situation

4. Assess how much time you have
5. Manage your team
6. Get details of potential witnesses
7. Take photographs and films
8. Make notes, draw diagrams, and use your voice recorder to keep a running record of what is there and what is going on
9. Preserve the evidence and take samples
10. Put together a basic idea of what happened, when it happened, and who was involved
11. Set up an investigation office

1. Securing the area

Securing the area is crucial. No-one should be allowed onto the site unless they really have to be there. The more people there are, the more evidence will be destroyed or disturbed, and the less safe the situation is likely to be. You need to keep as many people as possible out of the accident area until photographs have been taken and evidence assembled. Lock doors and gates where it is safe to do so making sure in particular that you do not create new hazards by blocking emergency escape routes. Put barriers or warning tape around areas as necessary and put up notices. Use security staff to control access. If the accident is at a remote site, you might not be there early enough to perform any of these functions. In that case, the planning and preparedness that you invested in earlier will be very worthwhile and the local supervisor or manager will already have started the investigation process.

2. Informing those who need to know

There may be legal requirements for you to inform regulatory bodies. The well-prepared investigator or HSE manager will already be aware of this and know how to do it. Telephone numbers will be on the mobile and in the diary. Managers and other company employees may need to be informed. The CEO will not be impressed if he finds out about the accident from a reporter ringing his door bell, or from the TV news.

3. An initial tour of the area

If you are there early enough, this is probably your only chance to see what the accident scene is like before it changes too drastically. Your on-site supervisor may already have taken photographs and secured important evidence, but you need to see for yourself as early as it is safe to do so and once you have dealt with the immediate priorities of making sure any injured people are safe and receiving attention. There is an Enquiry check-list at the end of the book which is useful as a reminder of what you might be looking for.

4. Assessing how much time you have

At this stage it is often useful to have an idea of how much time you have before attempts are made to return the scene to its normal function. Public traffic on roads can't be held up for too long. People will have to be readmitted to the area, and managers will want to get machines and processes up and running as soon as possible.

5. Managing the team

You may be working with a team, and it is good to have assistance even with a minor accident investigation. Gather your team together and organise and delegate so that they know what they are doing to help the investigation. You might need to bring in specialists, so ask someone to contact them and arrange for them to be brought into the investigation process. If there are complex systems, machines, or computer records, bring in an expert. They may not need to join the team for very long, giving just enough time to the task to complete their part of it.

6. Getting details of potential witnesses

We consider witnesses and interviews in detail later on. At this stage, get basic contact details from anyone who is or might be a witness. Anyone who might have seen anything, heard anything, or done anything, is potentially a witness. Ideally, have someone who is good at this assigned to your team so that they can concentrate on this aspect of the investigation.

7. Photographs and films

Films can be useful to give a general view of the site, while photographs can show more detail. Modern cameras enable the investigator to take hundreds of photographs, which can then be downloaded onto a computer and viewed instantly. This is very useful, but the sheer number of pictures can lead to confusion. One way of avoiding this is to make sure that the camera clock is set correctly, so that the date and time on each photograph is accurate. Then you can use a voice recorder to describe what you are photographing at what time. "It's 2.30 and I'm taking photos of the main workshop ... It's 2.40 and the pictures are of the machine in bay 4" on the voice recorder means that when you load the pictures into the computer you can sort them by time and subject.

Anything other than intelligent amateur use of the digital camera is beyond the scope of the average HSE manager and certainly of this book, but here are just a few points that might come in handy:

- In your tool kit, include a ruler with very large clear markings, which you can place alongside objects to show their size.
- Similarly, for a wider angle shot, you can include some familiar object with a standard size, to give the eye a baseline by which to judge the other objects. If appropriate, ask someone to stand in the photograph.
- If you want to photograph what someone saw or might have seen, take the shot at eye level.
- It is sometimes helpful to take different kinds of shots of the same object or scene if it is potentially an important piece of evidence. Black and white photography isn't much used these days except by experts – but it can show gradients of contrast better than a colour shot.
- Taking pictures from four sides of an object can capture a lot of information. It helps if each of these pictures is taken from roughly the same distance away from the subject and at the same height.

There are various specialised kinds of photography which may be used, for example microphotographs to show fatigue in metals. X-ray and infra-red photography can also be used effectively, but for these you will almost certainly have to bring in an expert.

8. Notes, plans and diagrams

Photos and films are great but they can be misleading when it comes to spaces, distances and sizes. This is where diagrams come in. If distances are important, measure them and prepare a quick sketch plan. It can always be redrawn or tidied up later if necessary. What is important, however, is to take measurements as accurately as possible.

Most information that you need can be recorded on your digital voice recorder, but you may well find it useful to make notes as well – a "things I want to remember to do" list is not much good buried somewhere on the recorder.

9. Preserving the evidence and taking samples

Some evidence will be hard to miss and will stay put. But there is much evidence that might be easily destroyed. In particular watch out for:

- Skid marks
- Impressions in soft ground, for example of tyres or footprints
- Spillage of liquid, powder and other substances
- Debris
- Broken glass
- Burn marks
- Dirt
- Soot

Do not allow anyone to start hosing down the area or wiping and cleaning up until you are sure that you have all the samples you need. Liquids may seep away or run into drainage channels, so gather samples as quickly as you can. If the liquid you are interested in has disappeared into the soil, take a soil sample.

Paint is often a useful clue to what has happened. Samples of paint can be taken by scraping flakes off the surface into an envelope. Any painted surface will leave minute traces of paint on any reasonably hard surface with which it is in contact, thought this may not be obvious by just looking at it. Surfaces where traces of paint are not obvious but may have been left can be protected until microscopic examination is undertaken. Similarly, if someone has been struck, for example by a moving vehicle, traces of their clothes or an imprint will be found on the point of impact. Samples of hair, blood, etc may also be present.

Some evidence might not be washed or wiped away, but might be destroyed as evidence if it is moved. This is why photographs and sketches of the accident scene should be taken as early as possible. This is another area where on-site staff can be trained, so that if necessary a photographic record is already established before the investigator arrives at the scene.

Take samples of anything that might be evidence, carefully labelling the container with a description of the sample, the location, time and date. It is sometimes useful to photograph the completed sample in its container, at the place where the sample was taken.

In some cases, and with specialised help, samples of blood, urine and breath can be taken. In cases where someone might have beeen drunk or under the influence of drugs, this sampling can be crucial.

Be aware too of hazardous substances which might or might not also be evidence, for example

- body fluids, blood, excrement
- hazardous chemicals
- sources of radiation

10. Putting together a basic idea of what happened

At this stage it will probably help if you put together a basic outline, with lists of who was involved as far as you know at this point. This is the basis for later analysis. You can change it as often as you like, but a good framework early on can help.

11. The investigation office

The position, size, layout and contents of the investigation office will depend on many factors, such as how many investigators and other personnel are going to be involved; how long the investigation is likely to take; how many witnesses are likely to be interviewed; and how much evidence and documentation will need to be gathered together in one place. Make an initial judgement on this and make sure that the investigation office is large enough, safe, quiet, and secure. Separate interview rooms are essential, and separate areas for the secure storage of large bits of evidence such as machine or vehicle parts might also be required.

Ensure that access is restricted to those who need to be there. If the room has to be cleaned or otherwise serviced, and if there are people coming and going, make sure that there is no security risk. Your investigation office may contain a lot of personal information, witness statements etc, and there may be valuable equipment there. It would be disastrous if you came in one morning to find that the computers and cameras had been stolen though of course you would have back-ups of all your data, wouldn't you? (Please make sure you do!)

4.3.3 Preparing for the investigation

Here are some tasks that might fit into a short preparation phase before the investigation proper begins:

1. Updating the list of persons who might be involved

Find out who was involved and add to your list, including contact details. Your list should include anyone actually involved in the accident and anyone who might be a witness. Start making appointments for interviews.

2. Briefing and liaison committee

If there are several agencies involved, set up a briefing and liaison meeting so that everyone's role can be explained and co-operation established.

3. Who else is going to be involved?

Find out who is getting involved or who is likely to be involved in the investigation process and make contact with them. Other agencies and individuals could include:

- Lawyers
- Union representatives
- Regulators
- Police services
- Fire services

- Local authorities
- Owners' representatives
- Local pressure groups

4. Securing electronic data

There are many possible electronic data sources that can potentially provide important pieces of evidence. Establish what they are, make sure that they are kept secure and make backup copies as soon as possible. Many machines these days are computer controlled or have elements that are linked to computers (Computer Aided Manufacturing etc). If this is relevant, ensure that information from the machines or processes is secure so that it can be used in the investigation. Data sources might include:

- CCTV recordings,
- computer logs,
- schedules, machine logs,
- machine data records, etc.

5. Securing documents and paper records

These might include:
- emails
- memos
- meeting minutes
- risk assessments
- method statements
- JSA reports
- safe systems of work
- job descriptions
- work permits

- codes of practice

Some of this will exist in hard copy, some as electronic documents, and most as both. As with electronic data from machines, you will need to know what is potentially important, keep it safe, and if necessary make back-up copies. Having the authority to take copies of whatever documents you need is essential, as we noted above. You do not want to be getting into all sorts of arguments about what you can and cannot access when you already have so many things on your mind.

6. The standard accident notice

When a Level 1 or Level 2 accident happens, everyone in the workplace will quickly know about it. People will probably feel shocked and frightened, and in the absence of any other information they will speculate about what actually happened and why. Then they will talk to family, friends and neighbours, and eventually they or their friends will talk to the press. By the time this happens, the reality of the event has probably been left far behind and fantasy and speculation will have taken over. For this reason alone, it makes sense to have a standard accident notice available for use. It need not be detailed, but would perhaps provide a brief description of what appears to have happened, with the names and conditions of those who were injured (if any) and details of what precautionary measures are in place. For example, a workshop or site could be declared closed to all personnel. The notice should tell people that an accident investigation is under way, and ask anyone who has any information that might be relevant to contact the investigator or the investigation team office. It is very important that the notice should be brief and factual. Speculation and surmise has no place here.

Quite apart from the practical value of this notice, it serves another important function. By giving sensible notice of the accident event, management is showing its commitment to the working community. And if "the working community" at this stage sounds like a vague concept, consider how important the safety culture is in management's attempts to manage risks. You cannot successfully ask workers to do all sorts of things, keep you informed about all sorts of things, and then, when there is a problem, clam up and claim that there is nothing wrong and it's all under control. That means, *we* are in control. To put it another way, management will do well to issue an Adult to Adult communication here. "Yes, there is a problem, and we are dealing with it according to our pre-arranged emergency and accident planning. If you have anything that could help us, please let us know." Anything else is likely to be patronising and perceived as a transaction from the Parent ego-states of management to the Child ego-states of the workers.

4.3.4 The investigation
Having done the preparation, now it remains to get on and do the job. Much of what you will be doing has already been dealt with above, but here is another short list:

- Interview witnesses
- Keep in contact and co-operate with other interested parties
- Make sure that the press or PR representative for the company is kept fully informed
- Use the briefing and liaison meeting to good effect
- Gather and store evidence
- Put the evidence together and prepare for the report

4.3.5 Cleaning up

At some stage the accident scene will be put back in order, unless the damage is so great that the site has to be cleared. So remember that this in itself is a potentially hazardous operation. Make sure that good risk assessments are done and any necessary information on the state of equipment, stability of buildings and presence of hazardous substances is available. The clean-up operation should then proceed, like any other operation, so that risks are managed.

Chapter 5: Interviews and Witnesses

5.1 Interviews

Witnesses and interviews are not entirely separable as subjects but we will start by looking at interviews and then consider witnesses in more detail.

After any adverse event you will need to get information from people who were involved in some way, and as part of the accident investigation process we will already have identified potential witnesses, people who may be able to tell us something useful for the investigation. But getting that information is not always easy.

When an accident happens, people within the organisation may be shocked, upset and even traumatised. They may also be scared, wondering what the accident investigation is going to find and if they will be blamed for something. So the first thing to do is to make sure that terms and conditions of the interview process are clear and are known to everyone who needs to know. The contracting process is most valuable here. The whole process should be designed to invite the witness into Adult, by focussing on thinking (rather than emotions) and the here-and-now problems of the interview process. If someone is too traumatised to access much Adult thinking, then it is not safe to interview them.

5.1.1 The contract
In agreeing a contract with the interviewee, the following should be in mind:

- There must be genuine agreement. If the interviewee is forced into saying yes to something, or feels that they have to agree because they are an employee, this is coercion, not a contract.
- The process will work only if it is from Adult to Adult.
- What needs to be agreed?
- If the interview is recorded, or notes are being taken, what is their status?
- Is recording or note-taking OK with the interviewee?
- Who will have access to information gained in the interview?
- How will information be used?
- What will happen to information once the investigation is over?
- What are the implications for the interviewee?

5.1.2 Blame

You may not yet know much about the causes of the accident, so technically it may not be appropriate to declare that no-one will be penalised as a result of the investigation. What happens if you uncover gross negligence, persistent abuse of the rules and procedures, or sabotage? At the same time it is important to establish that the investigation is not an attempt to apportion blame. The eventual accident report might uncover a series of events which might lead others such as management or the police and regulatory authorities to take action against an individual or individuals. But recommending such action is not the task of the accident investigation.

When something has gone wrong and someone has been injured or plant and equipment has been damaged it is all too easy to start looking for someone to blame. As soon as the cry goes up "It was his fault!" everyone else can relax. If HE or SHE or

THEY are to blame, then all we righteous people are in the clear. We can go into Critical Parent and enjoy ourselves getting worked up over all the inadequacies of that lot over there. There are so many examples of this happening, in real life and in books and films, that people are pretty much conditioned to it and will take avoiding action to ensure that they are not going to be blamed.

On one occasion when I was teaching, I asked for a volunteer to do a mock interview, to demonstrate how the interview process might be done. A course participant offered to do this, so the others gathered round and we set up the interview process. Once I had put the "witness" at ease I agreed a contract with him. Part of this included the statement that I was not looking for someone to blame, but wanted only to gather important information so we could all be safer in future. I then asked him to tell me, in his own words, what he had seen. At once his hands flew up in the air and he cried "I wasn't there!!" At first I and many of the participants thought he was joking, and we all had a good laugh. But he stopped us and said that, in all seriousness, this was the response he often got from employees. He knew that they had been there; they knew that they had been there; but such was their fear of being blamed that they simply denied it. "I wasn't there – I didn't see anything – you can't blame me." We all adopted this simple strategy when we were children, to keep safe and avoid blame, and most of us store that way of behaving in our minds, even if we are not aware of it. So, as adults now, we may still react to suspected blame with an immediate response of "It wasn't me!" This is a Child ego-state reaction in response to what the individual may see as an aggressive Parent request or criticism. We should aim to bring the witness back into Adult, to be in the "here and now" so that they can access their memory as accurately as possible.

5.1.3 Conducting an interview

You will need some standard details about the interviewee; name, address, telephone number etc, and also their place in the organisation if that applies. So you could use this process to set the scene and establish the relaxed and non-threatening atmosphere that you want. For most people it is pretty unthreatening to be asked their name and address, so this is a good place to start and can help the interviewee relax and begin to trust you.

Here are a few simple guidelines for a good interview:

- Allow enough time. If the interview is rushed it will be less effective.
- Provide pleasant surroundings. Make sure that the interviewee is reasonably comfortable.
- Ensure privacy. The interview room should be secluded and there should be no interruptions.
- Make sure telephone calls are taken by someone else and switch mobiles off.
- Interview only one person at a time.
- Welcome the interviewee. Check that you know who they are.
- Introduce yourself and explain what the investigation is aiming to achieve.
- Agree a clear contract with the interviewee.
- Make it clear that this is not a "blame" exercise.
- Acknowledge that the interviewee may have had a traumatic experience, especially if there has been serious injury or a fatality.
- As far as possible, put the interviewee at ease.

Chapter 5: Interviews and Witnesses

- Ask first about things familiar to them, like their job.
- Use terms that the interviewee is comfortable with.
- Invite the interviewee to use drawings and sketches if that will help.
- Ask open ended questions.
- Check what you have understood by repeating it back to the interviewee.
- At the end, thank the interviewee for their co-operation.
- Give them a telephone number or contact they can use if they think of anything else they want to tell you.
- Ask if they know anyone else who might have seen or heard something, or who would be useful to the investigation.

This may come into sharper focus if we turn this advice around, and describe what a really bad interview would be like. So, to achieve a really terrible result, follow these simple guidelines instead:

- Be in a hurry and do not allow enough time for the interview.
- Conduct the interview in unpleasant surroundings. Make sure that the interviewee is rather uncomfortable.
- Ensure lack of privacy. The interview area should be in a public space and there should be interruptions and disturbance. Leave and return to the room several times during the interview.

- Leave your mobile phone on and have the main investigation response phone on the desk in front of you. Answer all calls.
- Interview several people at once.
- Do not welcome the interviewees. Be vague about who they are and get their names wrong on several occasions.
- Do not introduce yourself or explain what the investigation is aiming to achieve.
- Do not agree a contract, so that there is no agreement on what is going to happen or what each of you expects.
- Make it clear that someone is going to get the blame for this accident.
- Ignore the possibility that these interviewees may have had a traumatic experience, especially if there has been serious injury or a fatality.
- Do not do anything to put the interviewees at ease.
- Cut the small talk and wade straight into the difficult questions.
- Use long words and technical expressions wherever possible.
- Ask closed questions.
- Don't bother to check that you have understood what the witness has said.
- At the end, do not thank the interviewees for their co-operation.
- Ignore any requests for a telephone number or contact they can use if they think of anything else they want to tell you.
- Don't bother to ask them about anyone else that might be useful to the investigation.

This appalling list may have helped raise the question of how to ask for information.

5.1.4 How to ask questions
Asking questions may seem like a simple exercise, but actually it is quite a skill. As with computers, what you put in will determine what you get out. Asking the wrong questions may give you the wrong information and leave you without a lot of material that would be useful.

Consider the following three questions:

1. "So you were in room 42 and heard a loud noise at about 10 o'clock?"
2. "Mr Smith crawled into the machine?"
3. "At what time did you realise that Mr Smith was in serious trouble and when did you turn the machine off?"

These questions are poor and probably confusing. In the first place they show the witness that you, the interviewer, already know (or think you know!) what happened. You already have a story in your mind, and it won't take the witness long to see what that story is. There is then the distinct possibility that the witness will, whether deliberately or not, agree with you even if they don't actually remember, or even if they saw or heard something different.

Each of these three questions could be addressed to the witness from the Adult ego-state; but they are most likely from Parent, and will most likely be heard as coming from Parent even if they are not. Or an ulterior transaction may be heard as well. Such transactions are likely to invite the witness to go into Child and agree with whatever is said. And if the witness is already feeling scared, their Child ego-state will say something like *"Just agree with whatever he says, and let's get out of here. After all, he thinks he knows it all anyway"*. Closed questions may also imply that the questioner doesn't actually want any information that might go against what he already thinks (the ulterior transaction might be *"I know all this already, so you needn't bother to think too hard"*). A questioner who asks closed questions will almost inevitably end up being told what they want to hear.

The first and third questions also contain more than one piece of information and may therefore confuse the witness. Confusion is a natural response and does not reflect badly on the mental agility of witnesses in general. I hate being asked a question with several different questions packed into it. How am I supposed to give a correct answer? And suppose one of these pieces of information is correct but the other not what they remember? Suppose the witness was in room 42 but did not hear a noise? Or that they were in room 45 and heard a scream at 9.30? Before they can begin to answer the question the witness will have to disentangle what they agree with and what they don't, both in their heads and with you, and by the time that has happened everyone may be getting confused. The second question contains only one piece of information but may well be asking for confirmation of something the witness doesn't actually know. Did they really see Mr Smith crawl into the machine or is that an idea being planted by the question? Perhaps Mr Smith ran into it, or fell into it, or was even pushed into it?

Open questions which ask for one thing at a time are, on the other hand, much easier to answer, and much more likely to succeed in putting the witness at ease. For example, once you have established that the witness does have something to say, you could ask a really open question like

"So please tell me what happened that morning?"

If the interviewer says this from Adult, in a clear and non-threatening tone of voice, the invitation to the witness is to respond from Adult and simply to say what they believe happened that morning. Unless there is some hidden fear that we don't know about yet, this approach is about as unthreatening as you can get. And you might get the response:

"Well, I was in room 42 doing some paperwork, and I heard what sounded like a sort of explosion. So I went into the corridor and couldn't see anything, so I walked through to the process room and saw a foot sticking out of the machine and that really frightened me, so I ran to the control panel and pushed the emergency stop button."

Now there are several things you can do. You might want more detail about some of this information, and could ask the witness to describe the noise in more detail, or describe where the foot was. Or you can ask for new information, like what the time was when they heard the noise, and whether they saw anyone else, and whether they knew who it was in the machine (in three separate questions, of course!). Or you could ask for the witness' opinion:

"What do you think caused the noise you heard?"

And at any point you can thank the witness and confirm what you have understood by repeating it:

"So, thank you for that. Now what I understand from what you've said is that you were in room 42 doing paperwork, and heard a noise, and that might have been caused by the machine door slamming shut. So you went into the corridor and you didn't see anything strange, so then you went in to the process room. You saw a foot sticking out of the machine there and ran up to the control panel and pushed the emergency stop button. Have I understood you correctly?"

There are several positive things about this process. In the first place, you show that you are taking the witness seriously. You have listened to what they said without twisting it up or adding bits of information. You have respected their version of events and understood it. And as they hear you recount this, they may suddenly remember something else, because they will probably be running through the events in their memory as they hear you recount the events. Then they can interrupt and say "Oh, actually, I do remember that there was a strange smell in the workshop, like, well like paint or something and I didn't know what that was..." and you can take it again from there. Also you have asked for their opinion, which shows respect for their intelligence and competence. This kind of questioning is respectful and factual, in other words it comes from Adult and invites the other to come from Adult too. And notice that by reflecting back what I think the witness has said and asking them if that is right, I am inviting them, in Adult, to assess my evidence and offer corrections. Most people will find this unthreatening. The implication is that we are in this together, we are both trying to understand what really happened.

As an investigator, you should not be caught up by gossip, malicious or otherwise, or in wild fantasies about what might have happened. If the interviewee has a theory, then listen to it with interest, and ask what evidence they have for it. If there is no evidence, the theory is probably worthless; but you can listen to some of it and make a judgement, as the theory may be based on years of experience and contain some helpful insights. Opinions may be useful, or they may be useless, and it is part of your skill as an investigator to know the difference.

5.1.5 Interviews at the scene and follow-up interviews
The witness who has some really useful evidence may provide even more if you listen to them at the scene of the accident. They may be able to visualise things more clearly and may remember details that they wouldn't recall in the interview situation, though this will depend to some extent on how much the scene of the accident has changed. After an explosion or major fire there may not be much accident scene left, and what is left may not be either safe or useful in jogging the memory. And if the events are both recent and traumatic, the witness may find revisiting the scene very uncomfortable. Again, it is part of your skill to judge whether an interview at the scene would be practical, ethical and useful.

You often don't have to get all the information from just one interview. More than one interview is indicated if

> ➢ you want to gather some initial information but the witness is upset. Go as far as you can within the bounds of compassion, and then if necessary invite the witness to come back at a later date.
> ➢ the information you gained from the first interview is, when you review it, internally inconsistent.

> the witness statement is at odds with what other witnesses have said.
> your time line (see below) introduces some conflicts which you want to resolve.
> you think the witness may have more information which perhaps they are not aware of. After interviewing several people you may have a better idea of questions you can ask to fill gaps in the story.

If the witness begins to suspect that you are checking up on what they said, they may begin to feel that you are accusing them, in effect, of lying. In this case they are likely to become hostile, acting from negative Adapted Child. Avoiding this is usually to be recommended, so make it clear that this second interview is to clear up a few points that you didn't understand, or to add to the overall picture now that you have had a chance to listen to a number of witnesses. If the witness is co-operative and eager to help, you could take them to see the time line and discuss the various problems with them, but best not to do this with a hostile witness.

5.1.6 Recording the interview
Not everyone agrees on this, but in my opinion it is a good idea to make a sound recording of all accident investigation interviews. There are many reasons for this, and they include the following:

- You can replay the recording to refresh your memory if you need to.
- You can replay the interview and listen to subtle nuances of expression, awkward pauses, hesitations, and so on, which may give you a feel for the reliability of the witness.

- If ever there is a dispute about what was said, the evidence is clearly available.
- If there are any subsequent accusations of bullying, harassment, incitement, bribery, or worse, then the recording is a witness to what actually did happen. It is your insurance against malicious accusations.
- The fact that a witness knows that the interview has been recorded makes any later accusations against you extremely unlikely.

A small recorder is unobtrusive and, once the interview starts, will probably be forgotten. But it is essential to make sure that you have the interviewee's permission to record the interview. You can explain why the recording is necessary by saying that it will help you to refresh your memory of important facts. It is probably not a good idea to go into the other reasons as outlined above! This is not necessary and may unsettle the witness – it is in any case very unlikely that the recording will be used as evidence to counter a malicious claim. Do, however, explain who will have access to the recording and what will happen to it after the investigation is over.

A video camera records even more information and is better from that point of view, but the equipment is more expensive; it is more difficult to get good results; the process is much more intrusive; and it is quite likely to induce a state of nervousness in the interviewee. For these reasons I have never used video recordings. But we should remember that there may be CCTV cameras in the area as part of the security system. The quality of the recordings is not likely to be much use to you, and it may be difficult for you to get access to them anyway. It is best to have these cameras turned off in any area when you are interviewing; if you can't arrange that, find somewhere else. Otherwise, you would have to inform the interviewee that a system over which you have no control is recording what they say and do. This will inspire confidence neither in you nor in the interview process.

Some people prefer to take notes during an interview but this is actually quite unsatisfactory for a number of reasons. In the first place you are writing things down at the same time as listening, and that is difficult, especially as you will be recording something that you have just heard while listening to something new. And of course, unless you are trained to do it you won't be able to take down everything that is said, so there is going to be a bias. As you listen to what is said you are having to choose which bits of the evidence are worth noting before you have the full story. And your notes are not evidence of what actually took place in the interview, because you could have rewritten them after the event. But if you do decide to take notes, then the same clarity is required as far as their use and eventual fate is concerned. Who will read these notes, and what will happen to them after the enquiry?

After the interview, you might decide that an interview statement which the witness can sign would be a good idea. You can then write a draft statement and, listening to the recording of the interview or referring to your notes, add in anything that you missed in the draft and which you think is important. You won't normally need a transcript showing exactly who said what, for a summary is much more useful. Once you are happy with the statement, the witness can be asked to comment on it and, if they agree, to sign it. If they don't agree, then the reasons for this will be interesting. Is the witness afraid that they will be called on to defend that statement in court? Are they afraid that you will catch them out in inaccuracies or lies? Are they just nervous about seeing what they said turned into a document, which is much more formal than talking in an interview? A meeting with the witness may make the cause of their anxiety clear and perhaps you can convince them that it is OK and they have nothing to fear. The witness' anxiety may be in their Child ego-state and you may be able to reassure them on what they are worried about. If their anxiety is about something in the here and now then you may be able to negotiate a solution in Adult.

Finally, a word of warning about the legal status of recordings, notes and statements. You do need to be clear about the legal status of these and if necessary you need to tell the interviewee what it is. This is likely to be different depending on the seriousness of the accident and whether external agencies such as the police are involved. There are different rules in different countries, so make sure that you know whether lawyers, the police or other enforcing bodies can require you to let them have access to the transcripts or recordings. Do not assure the witness that everything they say to you is confidential and then find that your notes or recordings have been taken away and used as public evidence.

5.2 Witnesses

Some writers use the word "witness" to include people, documents, objects, and so on, but in this book "witness" means a person. As in the discussion above, we use the word to mean *someone* who was involved at or near the site of the adverse event and who has direct information about that event which might be of help in an investigation. This will normally mean that they did something, or heard or saw something.

5.2.1 Getting witness information as soon as possible
It is essential to identify witnesses as quickly as practicable, and to interview them as soon as possible. This is because:

- Witnesses may be people who just happened to be there, and who won't normally be there again. Once they have gone from the scene, you may never find them again.
- Witnesses may be able to tell you about other people whom they knew to be there, and these may be useful witnesses too. It may be easier to find these people quickly.
- As time goes on witnesses will forget details.
- As time goes on, witnesses will probably talk about what happened, and as they do this they will "make sense" of what happened. This is good for them if they have been upset and shocked, but people change what they think, and that changes their memory of what they did, what they heard and what they saw.
- People will be affected by other information, especially by what other people say about what happened. Rough corners will be smoothed out in the witness' mind, and they may conclude that *"it must have happened like that"*.

Of course, there is a great deal going on in the aftermath of an accident, so if you can, have an assistant whose task is to find out who the witnesses are and get information from them. Obtain brief details initially:

- Name, address, phone number, email
- Place within the organisation if relevant
- What they did, saw, or heard

Even if it is quite short, an initial statement can be really useful, not only for the information it contains, but as a framework for a later and perhaps more detailed account. The witness assistant can easily record what the witness has to say on a small digital voice recorder. It is sometimes suggested that if there isn't time to interview all the witness immediately, and if this is practical, you should ask them to write an account of what they saw. In some cultures this will be much more difficult than in others. If the workforce is largely illiterate or unskilled, then asking them to write down what they know is neither practical nor encouraging for people who may be easily intimidated. In some cultures, people may be very able to write statements but are not going to do so because they will not want to commit themselves in writing. They may be afraid that lawyers and managers will get hold of their evidence and use it against them; and in the case of lawyers this fear may not be unfounded.

5.2.2 Witness reliability

Next let's consider the vexed question of the reliability of witnesses. This is a fascinating subject and a lot of research has been done on it. Without knowing all the theory, as an accident investigator you do need to know something about how far you can rely on what the witnesses say.

The first consideration is how much time has elapsed since the accident. As we saw above, various processes will be going on and these will dramatically change what the witness thinks they remember as they change what they think it all meant. In our frame of reference, we attempt to make sense of the world, using information that we already have. When something startling or upsetting happens, then that disturbs our frame of reference and we are uncomfortable. We may make a decision to change our frame of reference, but very often in such situations we worry away at this new and disturbing material and try to fit it into our existing frame of reference so that it no longer presents a threat. After an accident or disturbing incident, we may start to change what we thought we saw or felt or thought. We may begin to feel that what we saw (or heard etc) couldn't have happened like that, or wasn't important anyway.[46]

One effective way of resolving the problem is to talk to people and tell the story. As we do this, the contradictions may sort themselves out in our minds. The more we tell the story, the more we come to believe it, and as the story changes so we believe the changes too. As a teacher, I know this very well. To illustrate a point, I may have a good story, but as I tell it to different groups of students over the years the story changes, grows, and (I hope) gets better. Here is a story to illustrate the point about stories! I once heard a very good story illustrating an important point about how to cope with an unruly and difficult witness. The lecturer told this story as her own experience, and as far as I know it was. So when I started teaching similar subjects I too used the story. At first I told it as her story, but that is cumbersome and much of the dramatic effect is lost. So it wasn't long before I started telling it as though it were my experience. Certainly, it works much better – but now I have almost forgotten the fact that it didn't happen to me! After telling this story lots of times, I almost believe that was me in that situation. It has become, in some sense, part of my experience.

So witnesses may very well tell you what they thought happened, which means that they tell you what seems to make sense to them. This is not necessarily the same as what they saw. Such witnesses are not telling lies, but we will do well to really listen to what they say, and if there is any hint that they have revised their evidence to fit into a theory or previous conviction, then we can attempt to find out what the immediate thoughts and observations were. One way of doing this is to use some sort of role play – but this is not easy for the unpractised. Otherwise, careful questioning can reveal an earlier layer of perception. In particular, watch out for any suggestion that something "can't" have happened. Is that really true?

5.2.3 The hostile witness, the know-it-all and the expert
A hostile witness is someone who doesn't want to help, from whatever motive – or who even wants to harm the investigation. A witness may be hostile for all sorts of reasons, but the usual one is that they are frightened and protecting themselves or someone else. Or they may have been through an earlier accident investigation and found it a very unpleasant affair, so they are not going to help you out on this one. Or they may be angry at some real or imagined persecution from the employer, so they aren't feeling co-operative. A hostile witness may deliberately present false evidence in order to confuse the issue or get someone into trouble. It won't take you long to see that the witness is hostile, and once you realise this, you will probably have to use what they say with the greatest caution.

The know-it-all has been there, done that, and knows it better than you do. Some people have these beliefs about themselves prominently displayed. But since most of us don't like to be invited to feel ignorant all the time, know-it-all individuals often become socially isolated, and their grasp of reality is often poor. This makes them very poor witnesses. Their bluster comes from the Parent ego-state and they will certainly want to be in charge; but the real problem lies in the Child ego-state where there are likely to be feelings of social and intellectual inadequacy. Unfortunately, the know-it-all is going to want to come and help you with your investigation, probably by telling you all about what you are doing wrong and how they would (of course) do it better. The rule for dealing with such unwanted information is: stay firmly in Adult, concentrating on the task in hand. Keep inviting or pulling them back into Adult by asking them to set out, clearly and not in too much detail, what they saw, heard, etc, and what, in their view happened. When they start rambling off into irrelevant stuff, pull them back into Adult by asking for clear information from your Adult.

The real expert or consultant is not the same as a know-all. The expert is an expert because they actually do know far more about the chosen something than you do. If you are dealing with a complex technology, you may well need experts to tell you what they think, and to interpret some of the evidence. Hopefully, if you are interviewing an expert, you can give them an outline of where you are and ask them to tell you the rest. The expert will probably be good at transacting from Adult to Adult, imparting information as clearly as possible. Asking closed questions is even worse than usual here, because you may not know what questions you might need to ask to get the technical information you need.

5.2.4 Special care
Of course we need to bear in mind that close to an unpleasant event, witnesses may be upset and even traumatised. This presents a dilemma as you want their information as soon as possible but they may not be feeling very good. Sensitive judgement is called for.

Children present another particular situation where care is required. Make sure that you have the child's parents' permission to interview them, and never interview a child alone.

A sympathetic but neutral other person should be present at the interview, but should not take part. The point of these precautions is to protect yourself against accusations of bullying, or worse. It is generally not a good idea to have the child's parent or parents at the interview as this is likely to introduce all sorts of pressures that you may not even know about. The child may decide not to say lots of things if they fear that at home later they will be in trouble.

Having said all that, children can often be good witnesses, able to describe things in detail and without as much bias as an adult. But again, on the other side, children can also make things up for their own reasons, or for no reason at all, and they can be extraordinarily convincing. I vividly remember an interview where a seven year-old child gave the most detailed and convincing account concerning a particular man whom the child knew. Upon investigation, it turned out that this man had not been involved at all, and indeed was a hundred miles away from the scene of the supposed events when they happened.

5.2.5 Problems with witnesses
If the witness refuses to say anything what can you do? This is an interesting question with a boring answer. Legal systems differ widely, but it is likely that you are unable, practically and morally, to force the witness to tell you anything. So, if they refuse then you have to accept it. After a serious accident, if regulatory authorities and the police are brought in, then the situation may change.

When you suspect that someone is telling you a fabricated story, the interesting question is, why? Are they trying to protect themselves, or someone else, and if so is this because they or their friend really have done something for which punishment is to be expected or are they just frightened anyway? The Child ego-state of the witness may be panicking for no good reason connected to the case in hand, and just spinning a yarn. As we noted above, when we were children, many of us learned to stay safe when things went wrong by blaming someone else. "It wasn't me, honest, it was him." That strategy is still lurking in the Child ego-state of each of us, and even though we are not children, it can come out in moments of stress or threat. Or the witness may be quite deliberately trying to get someone else into trouble. It is wise to be very careful with any witness who starts blaming others.

There may be more sinister reasons why someone clams up or tells you lies. Perhaps there are all sorts of things going on in the background and you don't know about them. I once had the interesting experience of undertaking a survey of workers' attitudes to reporting adverse events in a large oil company. The management told me that almost all accidents were reported, and in an industrial process as hazardous as refining oil, this was very reassuring. But the workers told me a very different story. On one occasion, they said, a man had been injured quite badly when a valve shattered and a piece of metal sliced across the top of his head. The work team on site bundled him into one of their own cars and transported him to a hospital in a distant town. He phoned into work the next day with some story about a relative being seriously ill, and took a week's leave. When he returned to work he said he had been injured when kicked in the head during a football game. There was of course no accident investigation in this case, because the HSE department were not told about the accident – but what a complex set of lies the team concocted. And why did they do this? Because they were afraid that their supervisor would be angry with them for spoiling their safety record, and that they would lose their bonuses. So if someone starts telling you lies my advice is to play along with them as long as reasonably possible, for you may find out all sorts of interesting things. Confronting the lies can also give you some useful information, but needs to be handled carefully.

5.3 Being on the receiving end: accident and criminal investigations

So far we have mainly dealt with situations over which you, the HSE manager and accident investigator, have control. But for a moment let's reverse the situation and look at interviews and interrogation when it is you that is being interviewed. This will probably happen when there is a fatal accident because the regulatory authorities will want to interview everyone who was involved in the event or who had managerial responsibility for HSE.

The first thing that you need to do is get the contract clear. What authority do the investigators have? What do they expect of you and what are you agreeing to? Are the people conducting the accident investigation also conducting a criminal investigation? The answers to these questions will depend on where you are and which industry we are talking about, and it is impossible to be more specific here. The important point is that you need to know. And you need to know in advance, so you are not still wondering about it when police cars come screaming through the factory gates, lights flashing.

Let's take the United Kingdom as an example to illustrate these points. In the first place, the legal implications will be different depending on whether the interviews are taking place in England, Scotland, Wales or Northern Ireland. Then again, in some industries, such as the railways, there is a distinction between the accident investigators and the criminal investigators. The first group will be concerned only with gaining information which will be used to understand the causes of the accident and to improve the SMS. The second group will, as their name suggests, be looking for breaches of regulations and for evidence which could lead to a successful prosecution. Where there is a distinction like this, you can speak freely to the accident investigators knowing that what you say will not be used in evidence against you. This will not be the case with the criminal investigators.

In most cases, though, there is no distinction between an accident and a criminal investigation. Whatever evidence is discovered could be used to improve the SMS or to take the company or individuals to court. This introduces a very difficult conflict of interests. On the one hand, you and your company are, we presume, interested in finding out what went wrong so that the SMS can be improved and future accidents prevented. So you want to co-operate with investigators and give them the information they ask for. But, and it's a very big BUT, you don't want to end up being convicted of an offence and being fined or imprisoned. So you don't want to give anything away that might incriminate you. The powers of various agencies in different countries vary greatly; in some cases the investigator can do virtually anything, while in other situations their authority is strictly limited. Or the situation may change depending on whether a formal caution has been given, but even here it may not be clear whether information gathered before a formal caution has been made can be used in court. You need to know so get the contract straight.

It should be in everyone's best interests for everyone to co-operate, and of course you want to do that. But your response needs to be carefully considered. Employees should be informed of their rights and told what they have to do or do not have to do. It would be most inappropriate for the employer to tell employees what they should or should not say to an investigator, but the rules of the investigation should be clear to everyone involved. Take the vulnerability of the witness into account. Someone who was simply a witness to what happened is unlikely to be personally at risk from prosecution, but other people may indeed be personally at risk, either because of their position within the organisation or because of what they may or may not have done.

Beware of using "technical" sounding language without being clear what the implications are. Incredible though it may seem, in the UK a man was convicted of gross negligence because after an accident he voluntarily told police that he had previously conducted a "risk assessment". He was subsequently convicted of performing a negligent risk assessment. If he had not given this information, or if he had said that he did "a basic check of the area", he would almost certainly have been in the clear.[47] This certainly doesn't seem fair, but it is a good example of why you need to be careful.

You also need to know what the status of the information given is. For example, if an investigator simply asks a series of questions informally, can the information be used as evidence in court? And what about information given in a formal interview?

So here is a basic check-list that may help:

- What powers do the investigators have?
- In particular, can they compel someone to answer questions in an interview?
- What is the legal status of information given informally, or in formal interviews?
- Does a formal caution need to be given before information can be gathered as evidence for a court process?
- When someone is going to be interviewed, are they personally liable?

In 2003 an article was published in which a managing director, Brian Harris, wrote about a fatal accident and the aftermath. He was interviewed under caution, and in the article he lists the types of questions he was asked. His advice is to read them and consider how you would answer them.[48] This is good advice. If you are interviewed the questions may be different, but the information the regulator is looking for is likely to be the same. Notice how many of these questions are not concerned with whether there is an SMS or safety policy, but how you know that it is effective. We shall be thinking about this again in Chapter 8. Here are the questions:

- How is your company organised to implement its safety policy?
- What is your role in the company and what does it entail?
- What are your qualifications for this role?
- What training have you had for this particular role with regard to your health and safety responsibilities?
- What is your role in safety management?
- How do you discharge your safety accountability?

- What information do you receive?
- What do you do with it?
- How do you know that the information is valid?
- How do you know that company procedures are being followed?

Chapter 6: Making Sense of the Evidence

6.1 The essential timeline and how sticky notes can make a difference

In section 4.3 we went through the stages of an accident investigation and saw some of the tasks that might need to be done. At the end of the process the accident investigator will write a report, a subject discussed in the next chapter. One of the things that those who read the report will be looking for is an explanation of what happened, and an opinion on why it happened. In other words, the task of the accident investigator is to take the evidence that has been gathered and present it in a way that those who have not been involved in the investigation can understand and, if necessary, act upon.

You may remember advice in section 3.1.13 above to avoid stating opinions or interpreting evidence unless it is essential to do so. That is quite right at the beginning of the process, but in the report that is just what you have to do. In order to provide an account of the event, the report writer is going to have to interpret the evidence such as he has it, and tell a story. The story may be dry, or it might be very dramatic, but it has to have a beginning, a middle, and an end. It has to say, this happened, this didn't happen, and this is (probably) why that happened. Mr Smith did this, Mr Patel did that, Mrs Kumar said this, and the result was that this happened. The story does not exist except as a construction in the mind of the investigation team, and it exists there only because they have taken a lot of trouble to listen to, understand and interpret other people's stories. The witness statements are stories. They are not fact, but the reconstruction of reality within an existing frame of reference.

Chapter 6: Making Sense of the Evidence

How do we organise all the evidence into a coherent story? One effective way of doing this is to put a long sheet of paper on a wall and draw a time line (rolls of pale wallpaper are invaluable, or several sheets of smaller size paper can be taped together). Draw a long straight black line from left to right. Mark the adverse event near the centre of the timeline. At the beginning, on the left, ... well you might have to work back to that. Events after the adverse event are marked on the line to its right. Along the line at various points there are pieces of information, fragments of other stories, which the investigation has gained from interviewing witnesses and assessing evidence. At the top of the paper put approximate time markings, eg days or, more usually, hours. If a section of the timeline becomes too crowded you can use another timeline on a different piece of paper, in exactly the same way but with larger time intervals.

For many of these fragments of information, a time will have been established. Computer records and other electronic data may show exactly when an event occurred such as a machine starting or stopping or the existence of an error message. Witnesses may remember times exactly if they are part of a routine or if they noticed something odd and looked at their watch to check the time. Shifts or work patterns often start and stop at the same time of day or night, so there are broad frameworks within which to work. One suggestion here is to use a black marker pen to insert onto the timeline those events which are pretty much established. So, if a shift started at 8.00am just as it always did, and if there is agreement among witnesses that this was the case, then mark the beginning of the shift in black. Where there are several witnesses, choose a different colour of sticky note for each one, write their observations on these, in order, and as you complete each one, stick it on or near the time line at an appropriate point. For a particular set of evidence, use a particular colour of note or pen so that you can easily group them together when you look at them. In this way, you may easily see relationships between pieces of information that you won't have seen by reading through reports. It may be that one piece of information determines that others must have come before or after it. Similarly, you may be able to establish more exact timings when events and witness accounts are set out in this fashion. Sometimes we need to see things in diagram form for them to make sense.

Keep a list somewhere of which witness is represented by which colour, but don't leave that information lying around for anyone to see, and don't put names on the timeline itself. In this way you preserve the confidentiality of witness' reports.

Another great advantage of the time line is that it is there for all the investigation team to see. Everyone can look at it, individually and in a group meeting, and discuss possibilities, discrepancies, problems. Wrestling with problems of interpretation while looking at the wallpaper can provide excellent results. Questions can be written up in red; lines can be drawn between different events with comments and queries. And you can have lots of fun inventing new ways of adding and processing information.

Finally, another advantage is that if there discrepancies they become very obvious. Mr Singh says he arrived at the workshop at 14.15 but this conflicts with the evidence of the delivery driver, who says he delivered a load of boxes to Mr Singh in the stores at 14.00 hours, that they both unloaded them, and that the job took about ten minutes. Mr Singh says it takes 15 minutes to walk from the stores to the workshop. According to the delivery driver, then, Mr Singh can't have arrived at the workshop before 14.25 at the earliest. Now you can ask for more information. One member of the team can be detailed off to interview Mr Singh again, and another can check the timing with the delivery driver. Did Mr Singh perhaps sign a delivery note and was the time recorded? It might not have seemed important to ask that question when the driver was first telling his story, but now it is. Another team member could walk from the stores to the workshop with a stop-watch, and also ask around to see if there is shortcut. There has to be an answer somewhere, and by targeting the discrepancy, there is a good chance it will be found. As new evidence comes in, and as old evidence is refined or revised, the sticky label notes can be moved or re-written.

6.2 Using MORT

At this stage you may like to use some kind of root cause analysis. There are many systems available, and each industry seems to have its own tried and tested model.[49] However, many of these systems suffer from being over-complex for our purposes, and sometimes there is confusion between a system looking for probability and one being used to analyse what actually happened. Having said that, some sort of tree diagram can be enormously helpful at this stage. I recommend the use of the Management Oversight & Risk Tree (MORT) for a number of good reasons. Firstly it is free (see the section on Resources at the end of the book). Secondly, it comes with a book on how to use it. Thirdly, it is comprehensive.

On the *Accidents: Causes, Investigation and Prevention* course we spend time learning how to use MORT, and my experience is that when delegates first see it they are horrified. It is vast, and complex, and some of the terms don't seem to mean much, and there are all sorts of lines and strange boxes. But it isn't actually too bad once people understand some of it and start to use it. So I summarise here what you need to know to use MORT in a fairly simple adverse event investigation.

In the tree diagram, LTA means "Less Than Adequate". Whatever it refers to was there in the system, but for whatever reason it wasn't doing the job it was supposed to do. LTA could apply to anything from the management system to a machine guard or a loose screw.

Information in the tree is given in phrases, and these are included in shapes, which are not all the same:

- Anything in a circle is a basic fault event and doesn't go anywhere else.
- Anything in a rectangle is a general description of an oversight or fault, and goes further, so there will be other shapes like rectangles or circles below it.
- A rectangle with wavy top and bottom is an event that is normal in the system, something expected and known.
- A diamond shape is a dead end.
- Triangle shapes under boxes show that the sequence of the tree is transferred somewhere else.

These shapes are usually linked by lines showing relationship. Almost all of them are linked by a symbol that looks like a small space rocket (pointed at the top and curved at the bottom). This means OR, so it tells us that the phrases in the shapes linked through this box are alternatives. One or more inputs is required. At a few points, the symbol is different. It is pointed at the top and flat at the bottom so it looks like a bullet standing

Chapter 6: Making Sense of the Evidence

on its base. This means AND, and tells us that all the things linked to it apply at the same time.

Sometimes the chart gets too big in one direction so a section of it is split up and put somewhere else on the page, and you follow the number reference. This looks complicated, but is simply a way of keeping the system practical. So a number to follow simply says "There is no room here to put the next bit of the tree analysis in, so go to number XX and follow it on from there."

So, using this basic information, let's look at an example (please refer to the diagram on page 148). The top rectangle is the accident. Under this you see the symbol for AND, linked to three conditions. This means that each of these three conditions has to have been present. Suppose a worker called Farange is using an electrically-powered bench saw. Suppose the mechanical guard protecting the worker from the blade was faulty and broke while the machine was in use. Suppose that as a result Farange loses a finger. So we start with the accident and move down from there. In this case the potentially harmful energy was the rotation of the cutting blade; the vulnerable person was Farange the worker and the barrier was the machine guard. A "barrier" doesn't have to be a physical guard, but in this case it is. If the energy had not been there (the machine saw was not powered) or the worker was not there, or the barrier was adequate, no accident would have resulted. So the tree tells us that in order for the accident to happen there needs to be some energy, somebody or something in the way, and the barriers need to be LTA. Now we are invited to look at each of these elements.

Using MORT to investigate the accident, we would move down each of the three branches of the tree. First you would look at the controls and barriers which were, or were not, in place to prevent the transfer of harmful energy to the person. Secondly you would look at the vulnerable people or objects, in this case Farange and in particular his finger; and then at the potential harmful energy flow or condition, in this case a rotating blade.

Here we will look only at one, the "Controls and Barriers LTA" rectangle. Under this, linked by the OR symbol, we find two possibilities. So you ask yourself, was the barrier itself inadequate, or were the controls and process inadequate? It might be one, or the other, or both. You inspect the machine guard. It was worn and broken, and was certainly LTA. Later, you will return to this OR gate and follow the other route, looking at controls and processes, but for now let us follow the "Controls and Barriers LTA" route downwards.

Here we move down through another OR gate, so we have four possibilities. Firstly, the barrier could be *on* the energy source, for example an insulating sheath on an electric cable. Secondly the barrier could be *between* the energy source and the person or object at risk, for example a screen placed between a machine and the operator. Or the barrier could be *on* the target or operator in the form of PPE, as when a welder uses a visor to protect his eyes from the welding arc and gloves to protect his hands from the heat. Or, finally, the energy source could be *separated from* the object or person at risk by space or time, or indeed both. When my dentist takes an X-ray of my teeth she and her assistant leave the room and operate the machinery remotely, separating themselves by space. Some operations are timed so that they occur only when the operator has moved away and is no longer in the danger area, as for example in the case of robot machines in enclosures. This is separation by space and time.

Looking at the four possibilities, we go through and see which apply to our investigation. We have already decided that the barrier was LTA (otherwise there would have been no accident) so now the tree invites us to further define what that means. In our example where the unfortunate Farange loses a finger because the machine guard broke, the barrier was between the energy source and the person, so this is the box we follow down. Now you will see that under the first rectangle there is a further branch of the tree, with three different shapes. Under the other three rectangles there is a triangle with an arrow. Like a lot of MORT this looks far more difficult than it is, and again this is simply a way of managing the space available. Each triangle just tells us to use the branch under the first box for each of them. So whichever box we choose, we will use the same little bit of the tree. This saves having to draw up the same boxes under each of the four and making the tree unmanageably large.

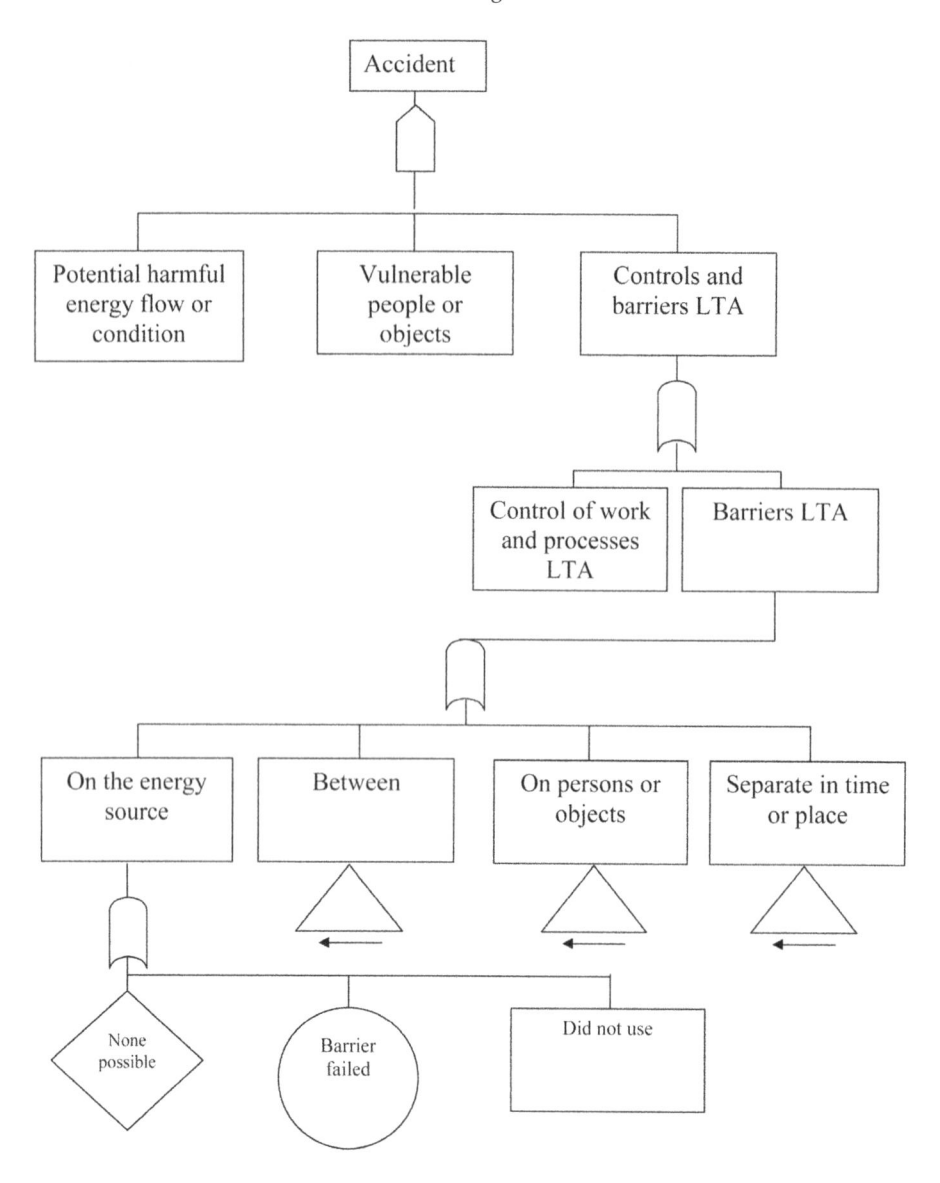

```
                    ┌──────────┐
                    │ Accident │
                    └──────────┘
                         │
     ┌───────────────────┼───────────────────┐
┌──────────────┐  ┌──────────────┐  ┌──────────────┐
│ Potential    │  │ Vulnerable   │  │ Controls and │
│ harmful      │  │ people or    │  │ barriers LTA │
│ energy flow  │  │ objects      │  │              │
│ or condition │  │              │  │              │
└──────────────┘  └──────────────┘  └──────────────┘
                                           │
                                    ┌──────┴──────┐
                         ┌──────────────┐  ┌──────────────┐
                         │ Control of   │  │ Barriers LTA │
                         │ work and     │  │              │
                         │ processes    │  │              │
                         │ LTA          │  │              │
                         └──────────────┘  └──────────────┘
                                                  │
              ┌───────────────┬───────────────────┼───────────────────┐
     ┌──────────────┐  ┌──────────────┐  ┌──────────────┐  ┌──────────────┐
     │ On the       │  │ Between      │  │ On persons   │  │ Separate in  │
     │ energy       │  │              │  │ or objects   │  │ time or place│
     │ source       │  │              │  │              │  │              │
     └──────────────┘  └──────────────┘  └──────────────┘  └──────────────┘
            │                △                △                △
      ┌─────┴─────┐
  ┌────────┐  ┌────────┐  ┌──────────┐
  │  None  │  │ Barrier│  │ Did not  │
  │possible│  │ failed │  │   use    │
  └────────┘  └────────┘  └──────────┘
```

The first possibility is that no barrier was possible. If that is the case, then we have reached a dead end, and so we have arrived at a root cause: "For operational reasons it was not possible to have a barrier between the energy source and the operator and this was a cause of the accident." The second possibility is that the barrier failed. Again, there is no further way down, so this counts as a cause. Finally, it might have been the case that the operator did not use the barrier. As you will see this is a rectangle, so the analysis goes further though I have not included it in my excerpt. Under here we are invited to consider whether a barrier was provided, or whether there was a "Task Performance Error". So perhaps a barrier was not provided, or perhaps it was provided but the worker did not use it.

Now this has probably not told you anything you couldn't have worked out for yourself without a complicated looking diagram. Sometimes students look at it in disbelief and wonder why we would want to use it! But there are indeed several good reasons. One might be that an analysis tree like this is really a kind of check-list, and by using it we are prompted to ask all the relevant questions. Certainly, you could simply devise your own check-list and tick off the items one by one, but it would perhaps be difficult to see the connections between the items. Secondly, if we devised our own list every time there was an adverse event to investigate, this would be most inefficient. It would take a long time and we would inevitably miss things out. MORT gives us a ready-made check-list to save effort and time. Once you have used it a few times, it becomes a really valuable friend. Also we should consider that accident investigation is not the process of looking at bits of what happened and saying, well, I think this is the cause. If you do that you will end up with the obvious. Using MORT gives you a structure. In our example we have looked at just one small corner of it. Used as a whole MORT invites you to look at all sorts of possible causes, for example, inspection, maintenance, supervision, support, technical information, specifications, the risk management system, stabilisation and restoration. Take a look at the full MORT and you will see how comprehensive it is.

The user's manual provides a whole series of questions to ask. As you use MORT to ask question after question, you will be directed to consider latent errors. The example we used just now concluded that the barrier failed. But why did it fail? What was the maintenance routine? If the problem was a failure caused by wear and tear, what sort of ageing was it? Why didn't the maintenance programme predict that this part was going to wear out in this way? Why didn't the worker report that the guard was old and broken? Were there supervisors and did they notice anything? If not, why not? Was the purchasing policy LTA? Was the worker properly trained? And so on.

One final point here. I agree with my colleague Rolly Angeles that "root cause is about learning from the things that go wrong"[50] and that this cannot be done in five minutes. Using MORT, even on a simple example like the failure of a machine guard, takes time and effort. But the beauty of it is that a fairly simple adverse event (we don't have to wait until Farange loses a finger) can lead to an investigation of the whole system and that can uncover all kinds of latent errors. Then you can sort them out and the potential savings in terms of money and HSE can be enormous.

Chapter 7: The Accident Report.

The accident report needs high status. How important are accident reports in your organisation? Who sees them and what is done with them? Do they have a clear route to the people who need to see them, or do they gather dust on a shelf somewhere? Are reports sent off into the void like orphans, never to be seen or heard from again? The status of accident reports will best be established in advance, before the first one is ever written. Before any adverse event has happened, management can set out to make everyone in the organisation aware of their real commitment to HSE and to accident investigation in particular. If it is known that management has set up a system and invested time and money and resources in accident investigation and risk control before any accident has caused losses, and entirely without pressure or demand from regulatory bodies, this is impressive indeed. In this chapter we are going to look at the practicalities of the accident report, and in Chapter 8 we shall find suggestions about how it can be used effectively.

7.1 Contents of the report

By the time you get to writing the accident report, you have put a lot of hard work and skill into understanding accident causation and theory, investigating the accident itself, collecting evidence, taking photographs, interviewing witnesses, developing a time line and all sorts of other things as well. Your accident report will, we hope, reflect all that. The report is the culmination of all your efforts, so let's make sure it is first rate. As elsewhere in this book, the lists and suggestions are to help you design your own report, and to develop a working method and a presentation that suits you best. Please use whatever information helps you and put the rest on one side as potential help in the future, or even as something that you really do not want to use at all.

It is perhaps important to remember that in writing an accident report, you are not presenting all the evidence. In a court of law, theoretically at least, all the evidence is presented and the judge or jury make a decision based on it and the arguments they have heard. This is not what you are doing, even though it may seem like it sometimes. You are presenting a report, which is something for others to read and accept. If there is a serious question about something, then the evidence can be presented in full and a discussion can follow.

7.1.1 Standard and individual accident reports
For many smaller and in-house investigations for Level 3 and Level 4 events, a standard form should be sufficient. The advantage of a standard form is that it acts as a check list and helps you remember what to include. Information on a standard form tends to be standard information, so it is easier to analyse. The disadvantage of all prepared forms is that they point you in a single direction and may lead you into a narrower understanding than you need. If that seems to be happening then you can expand the space available, perhaps by adding material in an appendix. There are some good worked examples of a standard report form available from the Health and Safety Executive free of charge, and there is also a blank form in the document, which you could download and use as it is, or adapt it to your own needs.[51]

For a Level 1 or Level 2 accident, or if there is some reason why a standard form will not do justice to the situation, then you will probably want to write an individual accident report.

First of all, it makes good sense to agree the contract or terms of reference at a very early stage. These will determine what goes into the report, and agreement should also be reached on who will read the report and how public all or some of it might be. As you complete your work, it is also a good idea to produce a draft report for people to comment on before the final version is produced. Again, who should comment on it will depend on the terms of reference, and on your common sense given the particular circumstances. But allowing others to comment on the draft must not be an opportunity for management, or anyone else for that matter, to censor your work.

7.1.2 Writing the report
The following guidelines may prove useful:

1. Introduction: set the scene by describing, very briefly,
    a. What happened
    b. When it happened
    c. Where it happened
    d. Immediate outcome (loss)
2. The terms of reference agreed by the team
3. Team members
4. Relationship to criminal investigation (if any)
5. Description of the scene, in more detail
    a. What was the process / activity?
    b. Who was involved?
    c. Where? Use photos, plans and diagrams
6. Objective description of accident
7. Evidence; what were your main sources of evidence?

      a. People: witness statements
      b. Things
      c. Documents
8. What everyone agrees happened
9. What we think happened
      a. Be clear about being unclear
      b. Give evidence for what you think
10. Conclusions
      a. Will depend on terms of reference
11. Recommendations for future action
12. Appendices
      a. Table of contents
      b. Index
      c. Terms of reference in full
      d. Members of team, details
      e. Documentary evidence
      f. Photographic evidence
      g. Acknowledgements
      h. References

Let us look at each of these in turn:

1. Setting the scene

This should be very short, and is here to give just enough information so that the reader sets up a framework of understanding in their consciousness.

2. Terms of Reference

Now the reader needs to know what your aims and objectives are. What has been agreed about the nature, scope and extent of your report?

Chapter 7: The Accident Report.

3. Team members

The reader should know who conducted the investigation. Who led it; who were the team members, and what was their position. In other words, why these people?

4. Criminal or other investigations

If other investigations were or are being conducted at the same time, then the reader should be told what they were and what the relationship was between the investigations. If certain sections of the report have had to be kept back until a criminal investigation is complete, then this can be noted.

5. More detailed description of the scene

Now the report proper can begin. Your aim is to give a complete description of the scene and what was happening before the adverse event so that when that is described the reader has all the information they need. Using diagrams and pictures can save an awful lot of difficult description. You can embed photographs in the text, or have them in an appendix, or both. Both is a good idea for a more complex report: put the essential pictures in the text and additional ones in the appendix. You might like to use the Enquiry check-list to help mark off important items in the report.

6. Objective description of the accident

That word "objective" is tricky. What you have done in this investigation is to gather together other people's stories and put them together, with your interpretations of the evidence, into your story. It may be the most plausible you can think of, and it may be the best explanation of the events that anyone is going to get. But it is not *fact*. Everything is interpretation. This concept sometimes seems strange to HSE managers as opposed to history graduates, so here is an example to illustrate what I mean. Imagine going into a large library and looking for books and articles on, say, Margaret Thatcher or Tiger Woods. There will be quite a lot to read: some of it will say how wonderful these people are; others will condemn and blame them. And there will be lots of opinions somewhere in-between and mixing both points of view. Thatcher and Woods are still alive, and could be asked about the facts of their lives; yet it is not the *facts* that are important when it comes to writing about them, it is the writer's opinion that matters. Two different authors can take the same set of historical information and produce two stories that are almost completely different. Ten authors, using the same information, will produce ten different accounts, ranging from adulation to contempt. There are bare facts, of course, such as dates of birth and when they got married and to whom, and what they had for breakfast on 10th July 2009 if that information is still available, but the only interesting part is the story, and that is interpretation. So it is with an accident report. We attempt to uncover the facts, and have a duty to be as objective as possible, but it is the story that counts.

Some of the report's interpretation comes from other people, some of it from the team and from you. What you present is the most objective account you can. So describe it in simple factual terms. Here is an example.[52] (Remember that all the relevant details outlined in the sections above will have already been given):

> "At 3.45 the two workmen reached the area where the test on the cooling fans was to take

place. Mr A went to the switch box, unlocked it, and switched off the electricity supply to a fan. He intended to switch off the supply to Fan No. 1, and was in no doubt that this is what he had done. After the accident, it was found that the power supply to Fan No.1 had not been turned off; but that the supply to Fan No.2 had been turned off. According to Mr A Mr B did not check to see if the correct fan had been isolated. Mr A informed Mr B that the supply to Fan No.1 had been turned off, believing this to be the case. Mr B then climbed up the step ladder and started to work on Fan No.1. About one minute later the automatic cut-in for the fan was activated by a temperature sensor. Since the electricity supply to the fan had not in fact been isolated, the fan started and Mr B was killed instantly. The witness statement from Mr A is in appendix 12e, along with a signed statement by the chief electrician at the site. Photographs of the switch box are in section 12f."

Notice that there are no explanations at this stage. We do not know why Mr A turned off the wrong switch, or even that he did. In theory, other scenarios are possible, though the report does not mention them and therefore presumably there is no evidence for them. For example, perhaps Mr A did isolate the correct fan, but when no-one was looking Mr C came along and switched the switches. I mention this only to emphasise the point that our story in the report is what we consider the most likely one, it is not the only one.[53] The evidence from his statement, from the electrician, and from the photographs of the switches suggests a sequence of actions, and the report details these as the most likely sequence of events.

7. Evidence

By now you have told the story, and probably referred to much of the evidence along the way. However it might be useful to summarise evidence at this stage, detailing who were the witnesses you most relied on; which physical objects provided the most significant information, and which documents were of most help. The details of all these, so far as is possible, should be in the appendices, so that should he wish to do so, the reader can at this point look at the evidence that has most importance for the report's conclusions.

8. What everyone agrees happened

Now this might sound like a re-write of section 6, but actually it isn't. That told the story. The evidence for that has been presented. Now the report should distinguish between the generally agreed bits of the story and the bits that are uncertain, conjecture or downright obscure. In this section therefore, the report should set out the events on which everyone agrees. Something like this, perhaps:

> "The following description is generally agreed and there is good evidence to support this account: Jim and Abdul were the only two workers near the press at the time of the accident. Pierre was working at the press, and had been there for an hour and forty minutes. He was an experienced and careful worker with a good safety record and employment history.

Chapter 7: The Accident Report.

Pierre was the only worker who used this press. At 15.22 or thereabouts Jim realised that the press was silent, which was unusual, and went round to the press room to see what was wrong. He found Pierre unconscious with his left arm trapped in the press. There was a lot of blood on the machine and the ground around it. Jim called for help and ran to the emergency alarm, which he activated. Help arrived within three minutes and Pierre was taken to hospital, where he died six hours later without regaining consciousness. Subsequent investigation established that the interlock guard was not working effectively, and when the press was activated after the accident it was found to be jamming on a loose piece of metal caught behind the guide channels."

9. What we think happened

In the previous section we told a story, based on the best evidence we have, and we invite the reader to safely agree with our story. We may not be correct in every detail, but generally speaking we are confident that this is what happened. But in this next section, we are being much more speculative and flagging the fact up for the reader. This is what we *think* happened, but the evidence may be contradictory, doubtful or missing entirely.

"We believe that Pierre realised that the press was sticking and instead of calling out the maintenance engineer attempted to clear the problem himself. He was not a trained maintenance engineer but had many years' experience of working on this kind of press, and it is possible that he had carried out minor repairs before. A toolbox was found in Pierre's cupboard near the press and a number of new bolts were found in the support frame at the back of the press. We did not find any record of maintenance work which would have replaced these bolts. We were also unable to establish why the interlock guard was not working properly. No fault had been reported, and it may be that Pierre had worked with the machine in this condition for some weeks. It is also possible that Pierre disabled the interlock guard in order to carry out repairs to the machine – either on this occasion or some time earlier."[54]

A note on style: this account may be a bit turgid, but it is carefully constructed in two ways. Firstly, the sentences are short. The reader can take in each bit of information as they read each sentence and don't have to hold long sentences in their minds before they reach the full stop. Secondly, use of pronouns is avoided if possible. References to "it" and "him" can make the meaning ambiguous. In most text this doesn't really matter, but here we are being very careful about the account and who did what when.

10. Conclusions

Most of what you have said has been said, so this is really a summary. Draw attention to anything that remains unresolved, and to anything else of importance that has not been sufficiently emphasised. There should be no new material included at this stage.

11. Recommendations for future action

For many this will be the most important section. Again, no new material should be introduced and the evidence for anything here should already have been presented, with references to further information in the appendices. What sort of recommendations should the report make? Here are some ideas:

- recommendations should come only from the evidence and information in the report.
- this means that the reasons why the recommendations are being made are clear and supported by evidence and reasoned interpretation.
- recommendations must be reasonable and practical in terms of cost and practicability.
- it always helps if recommendations are accompanied by a brief discussion or plan showing how they might be implemented.
- and a time scale should be given too.

Managers making decisions on what you suggest need to know how long it will take to do what you suggest and how much it will cost. You may not know in exact terms, but you can indicate for example that this is something that will not cost a lot and will lead to great improvements, or that this is something that might have some good effect but that will cost a great deal of money.

7.1.3 Using witness statements

Before reading this section, you might also like to look again at the section on the time line, at 6.1 above.

If you have just a few witness statements, it probably isn't too difficult to make sense of them, especially if there are no major contradictions. If there are a lot of reports, complex findings or contradictions, it might help to tabulate them on a chart or spread-sheet. This has the advantage of bringing clarity to the situation, but it also invites you to categorise diverse bits of information as the same. What I mean here is that if you are producing a table, it will not be difficult to put in data if witnesses X Y and Z all saw the same thing. But if your witness statements don't match up, you may be tempted to make them match up closely enough so that you can put them in the same column.

Here are some other ideas:

- Use different coloured sticky notes to represent each witness and write out a few words for each of their statements. For example, using green for Mrs Giri, you would write out "heard noise"; "ran to next room"; "found machine on fire"; "saw someone running from room" and you could put these onto a plan of the accident scene.

- Take a witness, or group of witnesses after they have given their statements individually, and show them the time line. Ask them to describe their experience using the time-line as a guide.

- Take the witnesses themselves to the scene, and ask them to describe what they saw and heard and map it onto the real location using large sheets of paper.

- Use other people to represent the witnesses, standing where they stood and describing what the witness said they saw or heard etc.

The advantage of using the witnesses themselves is that the information will be more accurate and the scene more realistic; the disadvantage is that the witnesses may be influenced by what the others say.

However you do it, you will need to boil it all down to something useful. In the report, you might say something like "According to the witnesses, the following happened:" and then summarise it. However, if there is confusion or contradiction this needs to be made clear. Your best analytical skills will be needed to give the most likely interpretation of what the witnesses revealed.

7.1.4 Who should read the report?

1. The legal problem

We have already looked at potential problems with confidentiality and legal access to documents. If there is a serious accident and the company writes an internal report, the external investigators may demand to see it. There are several possibilities here, but what you must not do is to write a report and then edit it to provide a "clean" version for the investigators! This is probably illegal and in any case liable to get you into a lot of trouble. It is best to start off knowing that if a report is written, then it probably has to be shared with the external investigators if they turn up. Set the terms of reference accordingly. Also, be aware that if evidence is quoted in the report then people may be required to defend that in court. Make sure that everyone is aware of this and prepared to do it.

In some countries it may be possible to use legal privilege to avoid having to produce the report in court. In other words you have your solicitor produce the report as a "privileged" document. This may solve some problems, but it certainly introduces others. If the legal chaps produce the report there is bound to be speculation about why you wanted to keep the information out of the court, and out of public view. What are you hiding? It is really not a good idea to give the external investigators the impression that you are guiltily hiding information from them. In any case, the confidential report could easily be leaked to the investigators or the press, and that could be worse still.

2. The media and press

As already discussed, you will need to consider what you release to the press and media. Too little and they will suspect a cover-up. Too much and you may suffer reputational damage.

3. Management

An accident report is almost certain to conclude that something was wrong in the safety management system and that changes need to be made. In this case, the report needs to reach those who can and will implement the necessary changes. This probably means managers, line managers, supervisors, foremen and etc. Management will want to use the information,

Chapter 7: The Accident Report.

conclusions and recommendations of the report in order to improve the SMS, with all the benefits that this brings; and this is the subject of the rest of the book.

Chapter 8: Preventing Accidents

In the previous seven chapters we have learned something about the history and causes of accidents, how to understand some relevant aspects of behaviour, and how to investigate accidents and write an effective report based on the evidence. All that should be useful and, I hope, interesting; but these things are the servants of our main aim, which is to keep people safe and well, and to prevent loss. Indeed accident investigation is an important part of any organisation's loss control system.[55] Put another way, the aim is to prevent adverse events, and many benefits of this were discussed in Chapter 3. The end product of the investigation process itself is the accident investigation report, which itself needs to become the starting point of another process in order to achieve the goal of saving money and saving lives. In this chapter I am going to suggest the use of a committee as a good way of ensuring that this important goal is achieved. Let us look then, first at the role of the Safety Committee (SC) and how it might be run, and then at how this can be vitally important in preventing adverse events.

8.1 The Safety Committee

8.1.1 Is it worth it?
If an organisation has a safety committee, what difference does it make? The members of the committee *must* ask themselves

- what do we intend to achieve?
- do we achieve it?
- how do we know that we achieve it?
- are there ways in which we could do it better?

I use the word "must" here because I feel strongly that unless a committee does some critical thinking about its role it is very likely to be a waste of time. Any organisation probably spends a lot of time and money on assessing productivity, quality, and profits and loss. Without this basic information it doesn't know what it is doing or whether it's making money. So why should it be different with a safety committee? People are spending their time at the meetings, and that costs money. There are minutes and reports to be written, and that costs money. Then people might actually read these minutes and reports – and that takes time too so it also costs money. Surely any organisation will want to know that this expenditure is worth-while and that the committee achieves something that was worth the cost.

Let's assume for the moment that the SC does know what it is doing and does achieve some positive outcomes. The next crucial question is, to whom does the SC report? Whoever it is, they need sufficient authority to make a difference. There is no point in having a brilliant SC which comes up with all sorts of wonderful ideas, which then never come to anything. In fact, this is a very dangerous situation because the organisation is likely to think that everything is under control because "the SC does all that"; and in fact nothing gets done.

8.1.2 Who should be on the committee?
An effective SC will have the members and representatives it needs. Here are some thoughts on who might be involved:

- Management needs to be involved and committed otherwise those with the authority to make decisions are isolated from the safety process.
- Workers need to be involved and committed because otherwise the safety effort can easily be seen as just

another management fad that has nothing to do with the real activity of the workplace.
- If there are different areas or departments in the workplace, someone should assess which need to be involved.
- If there are several work areas separated geographically, then a decision needs to be made about how to involve everyone. Should each workplace have its own SC? Should there be mini safety committees in each workplace reporting back to a big SC? The organisation needs to be as simple as possible but large enough to do the job.
- The shop-floor or work area is central, but there are other departments that might be represented too. For example:
  - medical and health care
  - logistics
  - purchasing
  - IT
  - supply
  - personnel

Bear in mind that all managers, however remote they might seem from the shop-floor or construction site accident, are making decisions that may introduce latent errors into the system. Involving them in the SC, and in the evaluation of the causes of accidents, can be a powerful preventive action. People who never thought very much about accident causation can find themselves thinking positively about how their own decisions might affect HSE.

Over the years I have asked many groups of workers "Who is responsible for Health and Safety?" and a popular answer is always "Everyone!" It's a good slogan, and true. But the problem is that if everyone is responsible, then there is no clear contract about who actually takes responsibility. Everyone can assume that someone else is dealing with that. It is a recipe for disaster. So getting every relevant person formally involved in the SC is a good way of ensuring that they know about and take responsibility for their role, and that they can contribute to the overall H&S system.

8.1.3 Contracts for an effective committee

An effective safety committee will have a good chairman (male or female), and will be run according to a known and agreed contract. Both in general terms, and at the start of a particular meeting, the wise chairman will decide what needs to be agreed with the committee members and then make a contract with them. For example, the chair could say "I would like to conclude this meeting at 4.00pm, which gives us two hours. In order to do that I will keep the discussions reasonably short, and if there is clearly something that needs more time we can hold it over to next week's meeting. Is that OK with everyone?" There may be reasons why that is not OK and these can be negotiated, but generally such a proposal will be accepted and everyone knows where they are. For regular and more formal meetings the contract can be provided as a document and everyone can be asked if they will agree to it. It is so much easier to deal with someone who habitually turns up late if they have agreed to be on time. What follows here are some suggestions for success in running a committee:

1. The chair of the meeting must be a competent chairman. It is important to note that the most senior person is not always the best chairman. On the other hand the chairman needs to be someone with authority.

2. There needs to be a deputy chairman, equally competent and of sufficient seniority. Ideally this deputy should have some role in running the meetings, otherwise they can lose touch and be seen as powerless.
3. An agenda must be prepared in advance and circulated (by email) in advance. Committee members should be expecting to take an active role in the meeting, and so they need to know in advance what is going to be discussed.
4. The chairman will make sure that important items are included on the agenda.
5. Significant items will not be accepted under Any Other Business. AOB is a useful slot at the end of the meeting to mop up any minor points that haven't been dealt with. But beware the manipulator who tries to introduce some difficult item for discussion at this point. Everyone is tired, no-one is prepared for the subject, and time is running out. The effective chairman will refuse to allow a discussion under these conditions, and suggest that the item is put on the agenda of the next meeting.
6. At the conclusion of a meeting, the minutes should be written up and circulated to all members as soon as possible. This means members can look at the record while the discussions are still fresh in their minds. And it also means that where action points are given, people will have as much time as possible to do something about them.
7. At the beginning of the next meeting, the minutes of the previous meeting will be introduced by the chairman. The action points will be checked and the person responsible will report on what action has been taken. If

the action point has not been dealt with this is recorded in the current meeting's minutes. The chairman will be aware of action points that are not being actioned; and deal with this as appropriate.
8. The meeting should if possible be held at the same time on the same day every week, fortnight or month. This means that people can put the dates in their diaries for a long time ahead. In any case a list of the next agreed meeting times and dates should be contained in each set of minutes.
9. The committee must start on time. If it doesn't, members will think, next time, that it doesn't matter if they are late ... and the effect will accumulate until eventually people don't bother any more.
10. The committee should last for two hours at the most. Any longer than that and people will get bored and regard it as too much of a commitment. A committee, if well run, should be able to get through the agenda in less than two hours. If it can't, the agenda was too long.
11. The people who need to be there need to be there. Nothing will reduce the effectiveness of this meeting quicker than people turning up to find that others are not there. Next time they won't bother either and very quickly the meeting will cease to function.
12. Where there are different departments or areas or sectors attending the meeting, it is essential that one of their representatives does attend. Otherwise they will feel left out and less enthusiastic about attending the next meeting.
13. Ideally, there should be two people nominated as representatives from each area, sector or department, so

that one of them is able to attend if the other can't. But it doesn't work if either of the representatives attends very infrequently: they both need to be up to date with what the committee is doing.
14. There must be a dedicated and skilled Meeting Secretary who records what is said and decided and who produces excellent minutes. The minutes
    a. will record who attends, who sends apologies, and who just doesn't turn up (not many of those, we hope)
    b. should not include a blow by blow account
    c. should record any discussions and the results
    d. should clearly describe and allocate action points.
15. If something goes badly wrong, the committee minutes may be used as evidence in legal proceedings or accident investigations, so they need to be clear and coherent.

8.1.4 The Safety Committee and Accident Investigation
As we said at the beginning of the chapter, there is little point in having a wonderful system for investigating and reporting on accidents if nothing happens as a result. Apart from anything else, workers and managers will rapidly lose interest in anything to do with HSE and accident investigations if they can see no results. So it is very important that someone has overall responsibility for making sure that

1. there is adequate preparation for accident investigation
2. the investigation process is monitored to ensure that it is effective
3. investigation reports are monitored to see if they are effective
4. reports include or result in recommendations

5. recommendations get a positive response.

It makes sense for a senior manager to have this responsibility, and if so then they should be a member of the SC. There are many advantages to having a named individual taking responsibly for this. The main one is that it is better to have one person who is accountable than a group of individuals who might or might not be accountable, and who in any case can always say "I thought those other people were responsible for that!" If the job doesn't get done, it's difficult to blame a committee. But if a named manager is responsible and accountable, and the job doesn't get done, the manager is going to be in trouble. So, we hope, they will make sure the job *is* done. Committees are great for continuity and for discussion and co-operation but we have to be careful to avoid the situation where everyone is "responsible" and no-one is accountable, for in this case the job is unlikely to be started, let alone completed.

Safety Committees may have a lot of work to do and they may meet infrequently. For them, accident investigation may seem like a small part of the workload, but this would be a pity. So in anything other than a small organisation, it makes very good sense to have a special committee to deal with the planning, implementation and outcomes of accident investigations.

8.2 The Accident Investigation and Prevention Committee

The AIPC can be a key player in the management of risks in an organisation. Its membership can be a lot more focussed than that of the SC, and can include the accident investigation team. Ideally it will be chaired by someone with experience and training in accident investigation, and if that person is not also a senior manager, then a senior manager with responsibility for accident investigation and control should be a member of the committee. As we noted above, the most senior person on the committee does not have to be the chair.

I suggest that the AIPC should focus on three chronological areas.

- Strategy. Planning before an accident happens
- Response. Immediate response investigation
- Follow up. Implementing recommendations and feedback into the SMS

1. Strategy

Membership of the team, terms of reference, and provision of resources are all essential before an accident happens. The AIPC should consider what physical resources it will need and what training provision needs to be made so that enough suitable individuals are available for the tasks ahead. If the team is going to use any of the techniques discussed in Chapter 6 these should be practiced in advance. It is no good turning up at the accident scene with a whole load of theory on bits of paper and finding out how to use them by trial and error on the job. The AIPC could set up exercises where the team or teams who will actually be called upon in an emergency are given a real or imaginary adverse event to deal with. They can go through the sequence of investigation and report writing and see what goes well and what they feel uncomfortable with. Training can then be targeted on the areas where it is needed. In this process, it will probably be a good idea to interview witnesses, even if this is role-play, and it is useful to do this in the presence of an observer or two who already have some expertise in accident investigation and who can report afterwards on what they observed. If the company has access to video recording equipment, now is the time to use it. Interviewers will learn an awful lot about their performance by watching a recording of themselves in action. Whatever methods you use, such training will not only increase the expertise of the team, but also contribute greatly to their confidence. And it can be fun too!

2. Response

The AIPC can plan what should happen when and if an accident occurs and an investigation is triggered. This will be different for every organisation, undertaking and site, and so should be carefully considered so that it is specific. Generic bits of paper which give vague information and general action plans will not be very useful in practice.

3. Follow-up

Once any investigation is completed, the AIPC should spend time asking what was good, what could have been better, and what went wrong! By learning from an actual event the Committee can vastly improve the accident investigation response next time round.

The AIPC should also have a procedure for making sure that the right people get the right information from the report, and then they should follow up what actually happens. If the report seems to be getting bogged down and little action seems to be happening, then the AIPC should have a plan for changing that. The higher up the management system the committee can reach, the more effective it will be.

An effective AIPC will go a long way towards ensuring that, in the event of an external or criminal investigation, you can show that you had a safety management system that works, that responsibilities were assigned, and that you knew that the system was working.

Now all this is a considerable investment of time, resources and money, but any manager who doubts that something along these lines is necessary should remember a woman called Cheryl Eckard. She made legal history recently when she was personally awarded $96 million, a sum to be paid by her former employer, GlaxoSmithKline. As a manager at a group company she became aware of serious problems at a manufacturing operation, and tried to alert higher levels of management to the violations. On one occasion she even phoned the company's chief executive, but he refused to take the call. Instead of finding out what was going wrong and doing something about it, management apparently side-lined her and eventually she was made redundant against her will. After an eight year legal battle, GlaxoSmithKline have paid $750 million to settle criminal and civil charges.[56] I am sure they could have found ways to spend that money on something rather more useful. If the company had had a Safety Committee or Accident Investigation and Prevention Committee with information about the violations, the power to ask the right questions, the authority to get things done and to call on management to act, GlaxoSmithKline would probably have saved themselves a great deal of money as well as the embarrassment of being quoted in books like this as an example of how not to manage adverse events.

8.3 What difference does your Safety Management System make?

All managers are familiar with some kind of SMS feedback loop, and we have of course just used it in 8.1.1 when considering how to judge the effectiveness of a committee. First set out your policy and decide how you are going to achieve it (PLAN); then organise and implement it (DO); then measure and assess how it is working (CHECK); then feed back what you have learned into the policy (ACT); and then go round the cycle again. The origin of this way of doing things was devised by Walter Shewhart in the 1950's in America. One of Shewhart's students was Dr W Edwards Deeming, and he revised the idea and took it to Japan. The rest, as they say, is history.

What Shewhart did that was so revolutionary, and so successful, was to invite managers to make improvements in the system itself, rather than trying to correct mistakes at random, as and where they were discovered. By working with the system, said Shewhart, we can continuously monitor and improve it.

Shewhart argued that in any system there are variations. In fact in a system that is working fairly smoothly, 85% of the random variations within it are common causes characteristic of the system.[57] In other words, there are variations, and they are inevitable. They are statistically normal, he said, and are a part of the system. They cannot be eliminated, so to maintain safety the system has to be designed to accommodate them. Similarly, Reason argues that slips, lapses and mistakes are an inherent part of human thoughts and actions. We cannot eliminate them. So, we should do our best to understand them and produce a system which prevents them from becoming critical.

This understanding can be applied to the technique of accident investigation. As we have seen, for years it was standard practice to identify the "unsafe act" and "unsafe condition" and blame the operator. But Shewhart and, much later, Reason, ask us to see the origins of accidents in the complexities of the system. Unlike Heinrich, who saw the immediate causes as "man error" and the unsafe act, Shewhart and Reason ask us to look at the system and the unsafe decisions that may be made about a wide range of activities. It is useful to look at unsafe acts and unsafe conditions; but only if we uncover root causes and latent errors can we effectively improve and continue to improve the system.

If this is our understanding of how accidents happen, then we will be examining the system when we investigate accidents. This is not always popular with managers and supervisors. An examination of the system asks, "who is responsible for what?" and this means management too – in fact it applies especially to managers and supervisors. I think that Heinrich's identification of the unsafe act as the cause of accidents has been very convenient for managers, who could point the finger at the worker and say "that fellow did something wrong, so the accident is his fault". But any serious adverse event analysis, having identified an immediate cause, continues by asking "So what was less than adequate in the system that allowed this accident to happen?" As Professor Andrews put it neatly,

> "Every accident, no matter how minor, is a failure of organisation."[58]

There are hundreds of versions of the basic management cycle; it is ubiquitous in quality and safety control. Indeed in virtually every sphere of human activity, including agreeing to go to out for a meal with your family, the Shewhart cycle is a very good way of getting things done, and then of finding out what we can learn from our experience so that next time it is better. At every stage the cycle is most effective if it is a contract. For example, "Let's agree on what we going to do and how we are going to get there." The details may be complex and the discussions difficult, but the principle is simple.

The investigation of adverse events is an essential part of the Shewhart cycle as it is the only really effective way of getting the essential information we need to feed back into the system and improve it. Vincoli confirms this:

> "there is well-documented evidence that high levels of accident risk reduction can be achieved by the average business through proper investigation and analysis of accident cause."[59]

The SMS works well only if there is feed-back to continually improve it. That feed-back comes from learning about what has gone wrong. And that means that we need to know about adverse events.

8.4 How to get everyone to report adverse events and why it is important

A successful company or undertaking will gain essential information about its SMS from the reporting of adverse events. Conversely, an unsuccessful undertaking will gain information about adverse events only when they cannot be hidden – and as we have seen, employees can be very skilled at covering up even quite serious accidents. While I agree with Vincoli that "Management must establish a *clear policy* which emphasizes the importance of reporting in the investigation process"[60] the difficult bit is not establishing the policy but getting people to take notice of it and do what it says.

8.4.1 It is no good just telling them what to do

We know enough by now to understand that there is no point in management coming over all Controlling Parent and insisting that all employees report adverse events. Those who are supposed to do this will not like being told what to do, and will probably allow themselves to be pulled into Adapted Child. They may be in positive Adapted Child, in which case they will probably do what they are told for a while until they get bored with it. But it is also likely that they will allow themselves to be pulled into negative Adapted Child and then they will vigorously resist these clumsy attempts to get them to do something which, really, they do not want to do. It is fairly obvious why workers do not want to report adverse events. Workers are often afraid of what may happen if they are found to be, or thought to be, doing something wrong. Since adverse events are clearly something that the management does not want, the initial reaction of a worker who is involved in one is likely to be to shut up and hope they get away with it. Their fears of what may otherwise happen may be realistic, or be

Child ego-state fantasies, or possibly both! My experience in this area suggests that when workers are afraid to speak up they justify this with some common reasons:

- we will get the blame
- the supervisor will be angry with us
- we will lose our bonuses
- we might be transferred to a different section away from our friends
- the other workers will resent us
- we might lose our jobs

It is interesting to see how these relate to child experiences (and note that I do mean the experiences of children). It is not unknown for parents to discipline children by blaming them, being angry with them, taking their goodies away (usually computer games and sweets rather than bonuses, though pocket-money can be a target), and sending them to bed or not allowing them to play with their friends. Children are also often sensitive to the charge of "telling tales". I am willing to bet that we all hold memories of these experiences, feelings and actions in our Child ego-states, in which case it is hardly surprising that workers are quite afraid of admitting "mistakes", and resistant to filling in a form to say "I got it wrong".

There are other reasons too, to do with our basic beliefs about ourselves. For example, I may believe that I have to get things right, and indeed I may believe that am only OK if I *do* get things right (and "things" can be almost anything). Asked to inform management that they "got it wrong", someone with this kind of basic belief about themselves is likely to resist strongly.

8.4.2 Bribery and corruption

If workers, managers, supervisors and indeed everybody in the organisation are going to report on adverse events, then they have to want to do so. There has to be some benefit for them. Having arrived at this insight, some people have said OK, then we will give people rewards for reporting adverse events. You can reward people with money in the form of an improved bonus, or you can praise them and put their name up on a board and give them a medal. In many workplaces I have passed by dusty shelves displaying annual Health and Safety Awards. Sometimes the reward strategy works to some extent, especially if the rewards are for teams or work-groups rather than for individuals. A spirit of pleasant competition always helps to get people keen on doing something. Some people at least will overcome their natural unwillingness to report adverse events if you reward them enough. But it doesn't take a genius to see that the major problem here is that people will start exaggerating, or even inventing, adverse events, in order to win the prize.

According to some experts, although management believe that reward schemes work, workers are often rather indifferent. Asked what the conditions are for a prize being awarded, workers often do not know. And there may be resentment if the reward is to a department, team or shift and notoriously unsafe workers get the same reward as those who are conscientious.[61]

Another problem is that it doesn't take a group of workers long to realise that winning the "Greatest Number of Adverse Events Reported this Month Award" doesn't look too good three months in a row. And a further disadvantage is that if the rewards are substantial people will cheat to get them, and if they are not substantial people will soon lose interest and go back to what they always used to do, which was to ignore all these management requests. What point are they anyway? We might also observe that an ulterior transaction from management might be, or might be heard to be *"The only motivation that matters is money."* This can lead to an unpleasant situation where workers will only respond to anything if they are paid for it. The same thing happens if you get children to do the washing up or persuade them to tidy their rooms by paying them. It won't take long before any request to do something is met with the response "How much are you going to pay me?" It's an expensive way of doing things.

8.5 Safety culture

When I first started teaching HSE, I found the bit on safety culture rather unconvincing. I wasn't at all sure that it was all that important, and I didn't really know how to get people enthusiastic about it. However, I am older and wiser (at least as far as safety culture goes) and now it is really important to my ideas about risk management. So what is a safety culture? The Chernobyl disaster in 1986 concentrated people's minds on the effect that the culture within a complex organisation has on safety, and the term "safety culture" was first used in a report on that accident. The definition of the Advisory Committee on the Safety of Nuclear Installations is very widely accepted:

> "The safety culture of an organisation is the product of individual and group values, attitudes, perceptions, competencies and patterns of behaviour that determine the commitment to, and the style and proficiency of, an organisation's health and safety management. Organisations with a positive safety culture are

characterised by communications founded on mutual trust, by shared perceptions of the importance of safety and by confidence in the efficacy of preventive measures."[62]

Wow. I wouldn't want to have to measure it. Actually, though, this is really what we have been saying all along. "Communications founded on mutual trust" are Adult to Adult transactions and probably involve contracts. "Shared perceptions of the importance of safety" means that everyone is involved and taking responsibility, and "confidence in the efficacy of preventive measures" means that everyone sees that adverse events are investigated and that things change for the better as a result.

In fact, measuring a safety culture is not difficult. When I am doing a safety assessment, then as soon as I walk into a factory, oil refinery, processing plant, building site, opera house or whatever it is, I can tell if it has a good safety culture. And I have heard other safety professionals say the same thing. You know at once if this is a safe place.

If that's true for me, then it is probably true for the workers and managers too. To different degrees, depending on status, training, experience and whether they care one way or the other, they too will know if where they are working is a place that manages risks effectively. And what people believe about that is going to make a huge difference to the way they behave. Here is my instant safety culture test:

1. Are there policies in place, are they good and are they up to date?
2. Do people do what the policies say they should do?

3. Is there an adverse event investigation process, and is it any good?

This list at least has the advantage of being very short.[63] And it works. Look at any place of work. Go and find a policy, for example a Health and Safety Management Policy. Is it covered in dust? Did the managing director last sign it in 1987? If so then the organisation's safety culture hardly exists. Take two examples from this policy and see what they say. These sayings have to be contractual actions that can be defined, measured and fulfilled. If all you can find is vague statements to the effect that "we will strive and try to do what we can to improve", then forget it. Such statements are not worth the paper they are printed on and the safety culture is probably a matter of producing nice-sounding HSE phrases to keep people happy.[64] But if you can find two meaningful statements, then go and see if they are happening. If the organisation claims that they will keep to all relevant legislation and regulations, then choose one you know about and see if they do. Finally, ask who is responsible for accident investigation. If there is someone, and if they can tell you what the pre-event planning is and what responses are in place if an accident happens, then that's a pretty good indicator of a good safety culture. If no-one knows and there is no planning in place, then the safety culture is poor.

And here is my even quicker safety culture test: Ask to see an accident report and, having read it, ask the management to demonstrate that the recommendations have been put into practice (assuming that they were good ones) and if not, what action did the management take?

What affects the safety climate? Neal and Griffin, who have reviewed the research on this, divide the important factors into two types: firstly, policies and procedures and secondly local work conditions and practices.[65] We can summarise their findings like this:

- 1. How far is management seen to give safety high priority, and how well do they communicate this and act on it?
- 2. How far do recruitment, selection, training, performance management etc. enhance safety?
- 3. How good are the systems for risk control and accident investigation; and how good are the policies and procedures?
- 4. Do Supervisors take H&S seriously and support those workers who also do their best?
- 5. Does the group communicate H&S issues well and provide support and back-up?
- 6. Is communication on H&S good between the relevant groups?
- 7. How dangerous is the work?
- 8. Are employees under too much pressure to work harder so that the workload is too great?

The original list uses rather complex language (the table is entitled "Proposed First-Order Dimensions of Safety Climate at the Organizational and Group Levels of Analysis"), so I hope I have done the research justice in translating it into something simpler. It seems to make good sense, however: if management is committed, communicates well; uses recruitment and other areas of the business to support H&S; and if there are good safety systems and accident investigation in place, workers will see that their health and their safety are taken seriously. If those in charge at the sharp end take H&S issues seriously and are supportive, and if the message gets through; and if risks are controlled as far as is reasonable, and workers are not under constant pressure to get the job done whatever the cost, then the workplace will be, and be seen to be, a pretty safe place. Other research revealed, not surprisingly, that "a training programme that taught supervisors to provide feedback about safety and to make safety an explicit performance goal produced a marked reduction in safety-related episodes and produced an improvement in safety climate".[66]

In section 5.3 I provided a list from Brian Harris. Here is another one, which is advice to the management of any organisation. Very good it is too:[67]

- Check that your H&S policy gives clear direction to the organisation
- Understand it and make sure everyone else understands it too
- Clarify accountability. Who is responsible for what?
- Everyone needs to have appropriate training so that they can fulfil their HSE responsibilities
- Make sure that what is supposed to happen does happen. How do you know that procedures are being followed?
- How do you demonstrate that you have not been negligent?

If the Safety Committee or the AIPC is sure that these things are being done, then it is doing a very good job and the organisation's safety culture is likely to be excellent.

8.5.1 Accident investigation is the key

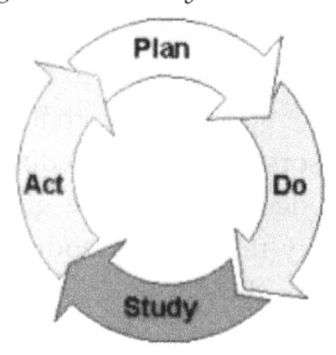

By now you should not be surprised to see the claim that accident investigation is the key to a good safety culture. If the accident investigation process is good then the safety culture is good. This is because:

- The management has taken HSE seriously
- Management has invested time, effort and money in HSE and the SMS
- The SMS can be improved every time there is an adverse event – and that means every day
- All the benefits of accident investigation become available to the organisation
- Members of the organisation will see continual improvement as recommendations are put into practice and the SMS is made more effective each time round.

If accident investigation is driving the continuous improvement model everyone in the organisation will feel positive. This is a large claim, and unlikely to be true in all cases. But if the management system is working, and people can see it working, their Child ego-state is likely to feel safe. The Adult information from the management is *"We take your well-being seriously and we are doing everything we know how to keep you safe and healthy. And, if you have any suggestions about how we can do this better, please tell us and we will listen to you. You are important."* And what about the Parent message? It might be *"We take your well-being seriously. There are things we have to do to make sure that you stay healthy and safe. And we are going to do that because you are important."* Now this is from Nurturing Parent, not Controlling Parent.

Doesn't that sound good? And, what is best of all, everyone can see that it really is happening! The Health and Safety Policy is not mouldering on a dusty shelf next to the risk assessments that haven't been looked at for years. When we look at the Policy and read that management says that it will keep the regulations that it has to keep (and even do better than that) we can look at the Welfare Regulations or the Use of Grinding Wheel Regulations and see that they are obeyed. Hazardous substances are treated with respect and protection is in place. The health of those working in the organisation is treated just as seriously as their safety. And if and when something goes wrong, there is a trained, experienced, competent team of people who will investigate, analyse and ensure that whatever lessons are to be learned will be learned – and the AIPC has the power and authority to make sure that the necessary changes are made, even if (and it probably will be the case) many of those changes are on the management side of things.

If this is how the accident investigation system works, are the workers going to report adverse incidents? If they are aware, in Adult, of the importance of this, and if they are encouraged to be a part of the process which is already keeping people safe and healthy, yes, they will. There is everything to gain and nothing to lose. This is not about admitting that *"I got it wrong"*. This is about "Here is something that I have seen, or done, that could be useful to all of us".

8.5.2 The importance of community

If the workplace is a community, people will feel loyalty to it. If management does their job properly, others will do theirs – and if they don't then management can get rid of them and find someone who will. If management is committed to creating a safe and healthy workplace which benefits everyone, then adverse events will be reported, violations will be few and the system will work, not only in terms of HSE but also in terms of excellent production and quality. On the other hand, if management are bullies who waffle on about all the wonderful things they are doing and don't then do them, and if management earns vast amounts of money while telling the workers that times are hard and they have to work harder for less, then don't expect the SMS to work with the co-operation of the workforce. Why should they co-operate? Would you?

Many of the things we have already talked about contribute to a good work community. Skilled contracting and firm Adult to Adult discussions and decision-making will encourage workers, supervisors and managers to treat each other with respect. This encourages loyalty. And trust. If the policies are clear and seen to be effective, trust and respect are again engendered. If there is a good safety culture and everyone really does take responsibility for HSE, then the sort of behaviour we saw Fred indulging in would simply not happen. Fred could discount the hazards in his situation only because it was acceptable for him to do so in the workplace culture. Presumably, whoever else was there on site with him took no notice because they all did that sort of thing.

8.6 The importance of feedback

Behaviour Based Safety works because it provides the right sort of feedback. Consider, for a moment, what you might say if you wanted to persuade a group of workers to wear ear defenders in a work area where the noise level is high enough to present a risk of long-term hearing damage. Well, you would say, if you don't wear this PPE you might go deaf in twenty years' time. And also, if I see you in this work area without your PPE I am going to be angry and, eventually, there will be disciplinary action.

Anyone with experience of HSE training will know that if this works at all it will work for a short time and after that things will go back to normal. Why? We have one model of understanding in ego-state theory. Your intervention comes from Parent, or is perceived to come from Parent, and pulls workers in to Child. They may do what you want for a time, but they are resentful and as soon as they can get away with it will revert to what they want to do – which is not to wear the PPE because its "sissy" and no-one else wears it, and it's uncomfortable, and we keep losing these ear-defenders anyway. And why *should* we do what we are told? So the way to solve this is to do it through contracting and positive feedback.[68]

A manager responsible for an increase in production, or someone who gets the work done on time and within budget is very likely to win praise and approval, and may be rewarded with money and promotion as well. All these are positive rewards, they are pretty certain, and they will happen immediately (praise and recognition) and in the medium term (promotion next year). On the other hand if this same manager starts to think about what could happen if something goes wrong because they are cutting corners and pushing the safety systems to their limit, then they are thinking about unpleasant events and negative results that might happen, at some indeterminate time in the future. Yes, there might be an accident now, but there might not, and since it hasn't happened yet, why should it happen now? So we have:

A	B
Positive	Negative
Immediate	In the future
Certain	Not certain

A shows us the feedback that the worker or manager gets if the job is done successfully (even if it is risky), and B shows the feedback associated with a possible accident. There is plenty of research that shows that most of us choose the feedback in A almost every time. Think of giving up smoking. Or not drinking too much. The immediate positive pleasure outweighs the possible negative effects that might happen in many years' time.[69]

It is exactly this calculation that makes it so difficult to get our workers to wear their ear defenders. The immediate feedback of *not* wearing them is comfort (because wearing them is not comfortable), it's immediate, and it's certain. There may be other positive feedback too, which is that since the safety culture is poor we don't get laughed at if we don't wear the wretched ear muffs. Not getting laughed at and teased is a powerful positive feedback for my behaviour in not wearing the muffs, and it's immediate and it's certain. When you try to persuade the workers to wear the PPE by telling them of some vague threat in the future, either going deaf or punishment by the supervisor, the feedback is very weak: it's negative, yes, but it's uncertain and it's a long way off.

So clearly what we need to do is give the workers some positive, immediate and certain feedback for wearing the ear muffs. This can be done through contracting and giving positive feedback to those who adopt the agreed behaviour. If there is a discussion about wearing this PPE, we may discover all sorts of problems, for example, that the muffs really are too big and heavy and uncomfortable, or that there are not enough to go round, or that they get lost. Solving these problems is an essential first step and shows that management is committed to its side of the contract. So, if we provide better muffs, and enough of them, and sensible storage, will you wear them? Eventually, unless things really are very bad in this workplace, a contract will be reached. Workers are now responding from Adult, not from Adapted Child, and this means they are more likely to take the long term possible ill effects seriously. It also means that they have agreed to do something, and so now it is (we hope) a matter of pride that they actually do it. Now the important thing is to give immediate, positive and certain feedback. So the supervisor gives praise to everyone who wears the muffs. This doesn't have to be overdone, but it does need to be consistent and immediate. The supervisor should also simply ignore those who are not wearing the earmuffs. After a while those who are still reluctant to wear the PPE will start doing so, because now *not* wearing the muffs is what gets disapproval or being ignored. And people don't like being ignored, they like approval. Behaviour Based Safety programmes are a more elaborate way of doing the same thing, and they work for the same reasons.

Good accident investigation is an important part of this process because it provides the basis for an ever-improving safety culture, which is what we need for this kind of contracting and positive feedback system to be most effective. And it also provides a model for the feedback process. It is extremely important that when workers report an adverse event, however large or small, there is an immediate, positive and certain response from management. If there is an accident report, in standard format or individually, there will be recommendations. It is absolutely essential that there is positive, certain and immediate feedback. This does not mean, of course, that management is immediately going to implement every recommendation in every accident report. Management is paid to manage, and that means that they have to make decisions about what can be done, in what framework and at what expense. But whatever the decisions, *these* must be fed back as a response. If the response is that the recommendations have been seriously considered by the Safety Committee and / or the AIPC and this and the next thing will be changed immediately; and the other recommendations kept for consideration in the refurbishment project next year; and that other recommendation is not practicable for the following reasons, that is a good outcome. Everyone has kept their side of the contract to do all that they reasonably can to keep the workplace healthy and safe for all concerned.

8.7 Accident statistics and how to make them work for change

Accident statistics are useful, but not very. Accident statistics are not reliable, and they only show you what happened when you were careless or unlucky. They are not a good guide to future performance. Accident statistics have to be produced and used by people who understand statistics, or action might be taken on the basis of figures that look significant but which are actually random variations within the system. In that case the actions will be irrelevant and a waste of money. On the grander scale, governments manipulate accident statistics and change the rules every so often so that comparisons are difficult to make. Of course in any workplace it is a good thing if your accident statistics show a statistically significant steady decrease in adverse events but the important thing is why they do that. Is it because your SMS is improving all the time? If so, well done. Or is it because you have fewer employees, or because certain statistics are not being gathered any more, or because people are increasingly frightened to tell anyone when something goes wrong? Or perhaps you are just lucky. Chance does play an important role here. Some people who are careful do still suffer loss; others who are reckless often get away with it.

My advice is to use accident statistics wisely, but don't rely on them too much. Other information is perhaps more useful. For example, letting everyone know that this year there were 24 tours of inspections of all the workshops, and on average the number of problems found was 36% less at the end of the year than at the beginning, might be more encouraging. When people learn that there were 52 adverse events reported during the first six months of the year, that 41 of them were investigated and as a result £300,000 was spent on upgrading and replacing machinery; and that in the second half of the year there were only 32 adverse events recorded and that production improved by 33%, then they will feel that they belong to an organisation that knows what it is doing and does it well.

Chapter 9: A Final Word

9.1 Don't go with the flow - keep swimming!

Some years ago I was part of a training group for people at a fairly high level of expertise. We would meet for a day, once a month. When we arrived we would set an agenda for learning and discussion. Some people came to the meeting well-prepared and would sometimes offer to talk about some area of expertise or some reading that they had found particularly valuable. Others had also thought about what they wanted and asked for a particular problem or topic to be discussed. But sometimes people had discounted the importance of the group and arrived with no ideas about what they wanted or were prepared to offer. When asked what they wanted from the day, they got into the habit of saying "Oh, I'll just go with the flow". This is an interesting transaction. At the Adult level it seems to be saying *"Well, you people are all very clued up and I'm happy to go along with that"*. For a long time no-one confronted the discount, which was in an ulterior transaction: *"This meeting isn't that important for me. I have been far too busy with other things and couldn't be bothered to make time to prepare anything"*. But one day the course convener simply said "The only thing that goes with the flow is a dead fish". After that, the phrase dropped to the bottom of the pond and was not heard again. People seemed to take more trouble in preparing to contribute to the group.

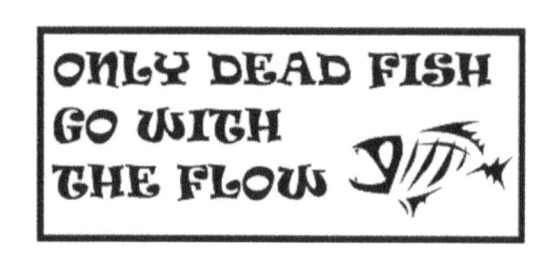

> ONLY DEAD FISH
> GO WITH
> THE FLOW

Chapter 9: A Final Word

James Reason uses the "safety space" model to show how safety can be a goal in real world systems, and how organisations can sometimes be like dead fish going with the flow. The safety space is shaped like this:

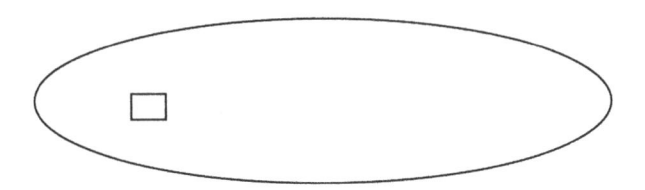

Within this area, organisations occupy a small space, like the rectangle in the drawing. On the left, organisations are increasingly resistant to operating hazards. On the right they are increasingly vulnerable. The organisation represented by the rectangle is doing pretty well. Organisations towards the left hand side will probably have a lower accident rate than those at the vulnerable end on the right, but the correlation is not exact. Chance plays a part. Organisations that "go with the flow" relax their efforts to maintain their SMS and accident investigations. After a period in which there are fewer accidents and things seem to be going all right, management may ask why they are spending all this money on HSE when there doesn't seem to be a problem. But there are treacherous currents which move the weak towards the vulnerable end of the space. If an organisation stops "swimming" it will drift passively to the vulnerable end of the safety space, where disaster lurks. Oddly enough, if serious accidents then begin to happen, the organisation wakes up and feels driven, by internal need and the pressure of regulators, to move in the other direction. The image in my mind (though this is not Reason's picture so I mustn't attribute to him) is that there a couple of sharks at the dangerous end that attack the passive organisations as they drift into snapping range. Faced with sudden danger, they renew their efforts and start swimming rapidly in the direction of the safer end of the space. Reason even asks the question "Do organisations need bad accidents in order to survive?"[70]

The safety space is another way of putting the complaint of my HSE manager friend Nick (see 3.2 above). They complain about paying you if nothing goes wrong; and then they complain about paying you when something does go wrong. In other words, you can't win. Move too far to the safe end of the safety space and things get relaxed, resources are withdrawn and the fish stops swimming. Move too far to the vulnerable end and disaster looms. It sometimes seems to be the case that in HSE we have to work hard to achieve nothing.

Or perhaps we could think of it another way. The HSE professional is like an experienced engineer lovingly maintaining and caring for the machinery for which he is responsible. He knows the sounds it makes, he knows how hot it runs, and can detect a problem by the sound and smell of the bearings. He knows when things are likely to go wrong and is ready for them. The engine that we maintain is the SMS. When things go wrong we know how to deal with them and to learn from mistakes so that the machine doesn't break down again. Every time we learn, the system gets better and the machine becomes more efficient, less prone to disaster, and more of a pleasure to work with.

We've moved from dead fish to purring machines. Whatever the image, we need good accident investigation to keep the organisation at the right place in the safety space; to keep the fish swimming so that it doesn't float hopelessly towards the sharks at the wrong end of that space; and to keep the engine purring along nicely, efficient, clean and a delight to work with. We have to keep on swimming. Here is Reason on the subject:

> "If someone tells you that they have a safe culture you will know to be deeply suspicious ... These are goals that have to be constantly striven for rather than achieved. It's the journey rather than the arrival that matters. Safety is a guerrilla war that you will probably lose (since entropy gets us all in the end), but you can still do the best you can."[71]

In describing the safety space, Reason defines it as "the achievement and maintenance of the maximum intrinsic resistance to operational hazards."[72] Accident investigation is the essential engine that drives that maintenance.

9.2 Saving lives and saving money

According to the Health and Safety Executive, 90% of all accidents are caused by human error and 70% of all accidents could be avoided if management took proactive action to prevent them.[73] The International Labour Organisation estimates that 2.3 million men and women die from work-related accidents and diseases every year.[74] If we use both these estimates, we can deduce that 1.61 million men and women could be saved from death each year if management took pro-active measures to prevent illness, disease and injury. We know that the numbers of accidents and injuries will be far more than the numbers of fatalities. So the numbers of humans disabled and injured, often with long-term and terrible results for them and their families, must be many times that, running into the tens of millions. That is every year.

Like the early mill owners, owners of factories and plants today often complain that they can't afford HSE. Once they start investing in all that stuff, they say, they will become uncompetitive and everyone will be out of a job. In the UK there is a powerful movement towards getting rid of much of the legislation that is designed to protect people in the workplace, on the grounds that it is too expensive. At the same time, the top managers see their income increase by enormous amounts every year.[75] The truth of the matter is that there is plenty of money to invest in saving money and lives. All we need is the understanding and the will to do it.

In a competitive market, it is increasingly the case that companies with poor HSE records are not selected for contracts. And no sensible manager wants to be part of a major accident investigation. But the problem with human beings, even managers, is that we are very poor at assessing risk. We are very good at assuming that "it won't happen to me". And we are also far more influenced in our choices by immediate gratification than by long-term risks. If we can do something that will result in positive feedback, quickly, then we are likely to do it.

Books on safety management systems and risk control often divide monitoring into two kinds: active (or proactive) and passive (or reactive).[76] Proactive monitoring involves thinking ahead, and managers might set up a system of audits, walk-throughs, and inspections, or use questionnaires and suggestion books to spot latent errors in the system before they cause problems. Accident investigation is presented as the main component of reactive monitoring, and we are told that when an accident happens then the system should react to it and learn from the mistakes that are made. I don't want to quibble with this, as it is a convenient way of looking at an important management concept, and the discussion in the Health and Safety Executive document (see footnote above) is excellent of its kind.

However, I want to suggest that accident investigation will work best when it is understood and used as the central and *proactive* element of the SMS. Certainly, when there is an adverse event, react to it and learn. But accident investigation is so much more than that, and by being planned and put into readiness long before a serious accident demands reaction, it can create an excellent safety culture and effectively maintain it. If this approach reduces the annual destruction of 1.61 million women and men and the millions more who suffer disablement and illness, even by a few percent, then it will have been a very worthwhile investment.

Resources

There are lots of good resources at http://www.rmlibrary.com/lib/safetymgt/accidents.php These resources are not free; membership is required.

MORT is available free at http://www.nri.eu.com/serv01.htm

See also The Investigation Process Research Resources Site at http://www.iprr.org/

There are many good free downloads at http://www.hse.gov.uk/ and there are also excellent books and booklets which can be bought. Recently a large number of popular booklets have been made available free of charge so this really is a very good place to keep among your favourites. In particular, HSG245 is excellent (see bibliography).

See also the IOSH website; there are some good books available here. www.iosh.co.uk

Contact
Dr James Thornhill can be contacted at: beechhillsafety@googlemail.com or on 0044 7919 123683.
Website: www.beechhillsafety.co.uk

Enquiry check-list

This is a check-list for items to enquire about and observe at the investigation stage. It can also be used at the report writing stage to check what needs to be included and perhaps to group information together.

- Location and purpose of plant / process / site
    - where is it?
    - how do people get there?
    - what are the normal ways in and ways out?
    - what was going on here?
    - and why was it going on?
- Traffic
    - what traffic was involved?
    - where was it?
    - who was driving?
- The environment
    - enclosed spaces
    - crowding
    - housekeeping
    - noise levels
    - temperature
    - radiation
    - ventilation
    - gasses, steams, vapours
    - lighting
        - windows and natural light
        - other lighting
    - weather
- Machinery and equipment
    - what was there?

- what was being used?
- to do what?
- how?
- by whom?
    - where were they in relation to one another?
    - what safety measures were in place?
    - inspection
        - what were the inspection schedules?
        - who was responsible?
        - who performed them?
    - details of maintenance
        - what were the maintenance schedules?
        - who was responsible?
        - who performed the maintenance?
        - was there preventive maintenance?
        - were there
            - random failures?
            - early stage failures?
            - age-related failures?
        - was there any maintenance on live or moving equipment?
    - design and maintenance of machinery and equipment: consider
        - ergonomic design
        - normal wear and tear
        - abnormal use and degradation
        - improper loading
        - lifting beyond safe limits
        - in the wrong place
    - what were purchasing policies?
    - was there defective machinery and equipment?
    - what were the fault notification systems?

Enquiry check-list

- o what were arrangements for withdrawal of faulty items?
- o what were arrangements for the repair/replacement of faulty items?
- Tools
    - o what was there?
        - what was being used?
        - to do what?
        - how?
        - by whom?
    - o what safety measures were in place?
    - o inspection
        - what were the inspection schedules?
        - who was responsible?
        - who performed them?
    - o details of maintenance
        - what were the maintenance schedules?
        - who was responsible?
        - who performed the maintenance?
        - was there preventive maintenance?
        - were there
            - random failures?
            - early stage failures?
            - age-related failures?
    - o design and maintenance of tools
        - ergonomic design
        - normal wear and tear
        - abnormal use and degradation
        - improper loading
    - o what were purchasing policies?
    - o were there defective tools?
    - o what were the fault notification systems?

- o what were arrangements for withdrawal of faulty items?
- o what were arrangements for the repair/replacement of faulty items?
- Energy sources
    - o what were they?
        - electricity
        - gas
        - water
        - compressed air
        - other pressurised substances
        - gas cylinders
        - hydraulics
        - vacuum
        - radioactive substances
        - gravity
    - o how were they being used?
    - o who was using them?
    - o were there any known problems?
- Flame and heat
    - o open flame
    - o enclosed flame
    - o other source of heat
- What substances were being used or were present?
    - oils, lubricants
    - gasses
    - compressed air
    - chemicals
    - hydrocarbons
    - known hazardous substances
    - o was labelling LTA?
    - o MSDS available?
        - flammable dusts

- Who was there?
  - why?
  - what were they doing?
  - how long had they been there?
  - Is there anything about the people there that is relevant?
    - employment history
    - age
    - disability
    - sickness
    - stress
    - depression
    - fatigue
    - personal / domestic problems
    - motivation
    - training levels
    - experience
  - evidence of unsafe practices
    - operating without authority
    - operating at improper speed
    - violation of rules and procedures
    - failure to warn of problems
    - using equipment or tools improperly
    - using unsafe or failed tools or equipment
    - fooling around (horseplay)
    - use of alcohol or drugs
    - use of prescribed medicines
- PPE
  - not available?
    - why?
  - using the wrong PPE
  - broken, poorly maintained
  - training LTA

- - supervision LTA
- Who was in charge?
  - what were the levels of supervision?
  - what was the command structure?
  - was it operative?
  - were important people absent?
  - were there stand-ins?
- What were the normal routines?
  - work patterns
  - times
  - refreshment breaks
  - visits to
    - lavatories
    - washrooms
    - canteens
    - rest areas
  - prayer times
    - prayer rooms
    - chapels
- What risk control measures were in place?
  - permits to work
  - systems of work
  - procedures
  - risk assessments
  - method statements
  - previous adverse event history
- Medical and health care
  - medical services
  - health surveillance
  - liaison with local services
- Emergency systems
  - emergency evacuation
  - signs

Enquiry check-list

- warnings
- drills
- training
- liaison with emergency services

Bibliography

Angeles, Rolly, *World Class Maintenance Management: The 12 Disciplines* (2009 centralbooks, Quezon City, Philippines)

Barling, Julian and Frone, Michael, eds., *The Psychology of Workplace Safety* (2004 American Psychological Association, Washington)

Eiser, J Richard, *Social Psychology: Attitudes, cognition and social behaviour* (1986 Cambridge University Press, Cambridge)

Ferry, Ted, *Modern Accident Investigation and Analysis* (1988² John Wiley and Sons, Inc., New York)

Gunningham, N, and Johnstone, R, *Regulating Workplace Safety: Systems and Sanctions* (1999 Oxford University Press, Oxford)

Hall, Stanley, *Danger Signals: An Investigation into Modern Railway Accidents* (1987 Ian Allen Ltd, London)

Harris, Brian, "Directors' and engineers' responsibilities for safety - a cautionary tale" *Loss Prevention Bulletin* 172, pp4-9 (2003 Institution of Chemical Engineers)

Holt, Allan St John, and Allen, Jim, *Principles of Health and Safety at Work* (2009⁸ IOSH, Wigston)

HSE, *Investigating accidents and incidents: A workbook for employers, unions, safety representatives and safety professionals* HSG 245 (2004² Health and Safety Executive, London)

HSE, *Successful health and safety management* HSG 65 (1997² Health and Safety Executive, London)

HSE, *A guide to the Control of Major Accident Hazards Regulations 1999 (as amended)* L111 (2006² Health and Safety Executive, London)

Hughes, Phil and Ferrett, Ed, *Introduction to Health and Safety at Work* (2009⁴ Butterworth-Heinemann, Oxford)

Hunt, Tristram, *Building Jerusalem: The Rise and Fall of the Victorian City* (2004 Weidenfeld & Nicolson, London)

ILO, *Providing safe and healthy workplaces for both women and men* (2010 International Labour Office, Geneva)

IET, *Safety Culture* (2010 The Institution of Engineering and Technology, Stevenage)

Krause, Thomas; Hidley, John; and Hodson, Stanley, *The Behavior-Based Safety Process,* (1990 Van Nostrand Reinhold, New York)

Manuele, Fred A., *On the Practice of Safety,* (1997² John Wiley and Sons Inc, New York)

Neal, Andrew, and Griffin, Mark, "Safety Climate and Safety at Work" in Barling and Frone, *Psychology of Workplace Safety*, pp.15-34.

Owen, D, *Air Accident Investigation: New Edition* (2001 Haynes Publishing, Sparkford)

Radford, John, and Govier, Ernest, *A Textbook of Psychology*, (1991² Routledge, London)

Reason, James, *Human Error* (1990 Cambridge University Press, Cambridge)

Reason, James, *Managing the Risks of Organizational Accidents* (1997 Ashgate Publishing Ltd, Aldershot)

Reason, James, *The Human Contribution: Unsafe Acts, Accidents and Heroic Recoveries* (2008 Ashgate Publishing Ltd, Farnham)

Ridley, John, and Channing, John, *Safety at Work* (2003 Elsevier, Oxford)

Slapper, Garry, and Tombs, Steve, *Corporate Crime* (1999 Pearson Education, Harlow)

Stewart, Ian, and Joines, Vann, *TA Today: A New Introduction to Transactional Analysis* (1987 Lifespace Publishing, Nottingham and Chapel Hill)

Sutherland, Valerie; Makin, Peter; and Cox, Charles, *The Management of Safety: The behavioural approach to changing organisations* (2000 Sage Publications, London)

Tilney, Tony, *Dictionary of Transactional Analysis* (1998 Whurr Publishers, London)

Vincoli, Jeffrey, *Basic guide to System Safety* (1993 Van Nostrand Reinhold, New York)

Vincoli, Jeffrey, *Basic guide to accident investigation and loss control* (1994 John Wiley and Sons, New York)

Index

accident
   definition, 13, 39, 40, 42
accident investigation
   investment in, 151
accident pyramid. *See* safety triangle
accident reports
   need status, 151
   terms of reference, 153
   writing it, 152
accident triangle. *See* safety triangle
Accidents: Causes, Investigation and Prevention, 2, 9, 143
accountants, 19
adverse event, 42, 151
   definition, 13, 41, 42
   levels of, 90
Advisory Committee on the Safety of Nuclear Installations, 183
Andrews, Prof K, 179
Angeles, Rolly, 150
asbestos, 28
aviation
   safety of, 22
behaviour, 43, 44, 47, 48, 49, 50, 51, 52, 54, 55, 56, 58, 60, 62, 68
Behaviour Based Safety, 25, 67, 72, 191, 193

Berne, Eric, 44, 56, 57, 58
Bhopal, 21, 81
Bird, Frank, 27
BP, 84
Buncefield, 29
CCTV, 125
Challenger, 21
Chernobyl, 21, 22, 33, 183
children, 132
cleaning staff, 78
community, 190
compensation, 14, 16, 17, 25, 86
ConocoPhillips, 27
consultants, 9, 87, 94, 103
contracts, 153, 185
   volenti non fit injuria, 16, 17
customers, 85
Deeming, W Edwards, 177
Deepwater Horizon, 84
Eckard, Cheryl, 176
error, 24, 28, 29, 30, 31, 38, 40
factory owners, 15, 24, 200
Ferry, Ted, 41
Ford motor company
   Pinto model, 19
Fukushima, 21
general public, 79, 85
GlaxoSmithKline, 176
Hammurabi, 14
Harris, Brian, 138, 187

Index

hazardous substances, 28
hazards, 43
Health and Safety Executive, 12, 41, 42, 152, 199, 202
Heinrich, H. W., 18, 19, 23, 24, 25, 26, 27, 37, 87, 178
   accident causation theory, 18
   domino theory, 18, 23, 36
hindsight fallacy, 33
horseplay, 49
Iacocca, Lee, 20
incident
   definition, 13, 41, 42
industrial society
   development in UK, 14
insurance, 86, 89
International Labour Organisation, 199
IOSH, 202
Japan, 177
Japan Airlines, 21
Job Safety Analysis, 30
jokes, 59
Kegworth air disaster, 31
Kurzman, D, 81
lapses, 29, 30, 31, 33, 178
latent errors, 34, 37, 38, 39, 40, 76, 92, 150, 168, 178, 200
law, 15, 16, 17, 21, 86, 133, 135, 163, 200
   function of, 16
   legal privilege, 164

status of interviews, 127
status of minutes, 172
lies, 133
logistics, 168
lone workers, 79
loss, 151
   definition, 13, 42
machines
   safety guards, 32, 40
maintenance, 30, 75, 78, 149, 150, 159, 205
Manuele, Fred, 23, 41, 94
media, 83, 85, 91, 110, 111, 164
medical and health care, 168
Merkel, Angela, 50
mistakes, 29, 30, 31, 33, 177, 178
MORT, 174, 202
Neal and Griffin, 185
near miss, 41
   definition, 13, 41, 42
negligence, 17, 114
nuclear power, 21, 22, 33
operating procedures, 37, 38
paint, 106
peer pressure, 54
Personal Protective Equipment, 14, 32, 43, 73, 146, 191
personnel, 168
photographs, 103, 104, 106
press release, 84
psychology, 29, 37, 43, 44

purchasing policy, 37, 38, 39, 150
Reason, Prof James, 29, 32, 34, 178, 197, 198, 199, 218
   Swiss cheese model, 34
recruitment, 38
regulators, 151
research, 82
risk assessment, 21, 37, 47, 92, 112, 137
road traffic
   risks of, 22
root cause analysis, 143
sabotage, 33, 114
safe systems of work, 22
safety, 26
safety culture, 33, 54, 56, 69, 72, 82, 111, 183, 185
safety management system, 78, 82, 89, 136, 138, 164, 177, 179, 180, 188, 190, 195, 197, 198, 200, 201
safety space, 197
safety triangle, 27, 28
shareholders, 85
Shewhart, Walter, 177, 179
slips, 29, 30, 33, 178

slips, trips and falls, 71
society
   attitude to accidents, 21
specialists. *See* consultants
statistics, 26, 27, 28, 29, 92, 178, 194, 218
Stewart and Joines, 66
story, 130, 140, 155
Sutherland *et al*, 25, 218
Thatcher, Margaret, 156
training, 2, 37, 38, 81, 138, 150, 174, 186, 196
Transactional Analysis, 10, 44, 46, 50, 56, 57, 67, 68, 69
undesired circumstance
   definition, 13, 41, 42
unsafe act, 18, 23, 24, 25, 26, 31, 35, 37, 38, 39, 87, 178
unsafe condition, 18
Vincoli, Jeffrey, 20, 83, 179, 180
violations, 32, 82
voice recorder, 103, 129
witness
   definition of, 127
Woods, Tiger, 156

Footnotes

[1] Most books on accident investigation are expensive, even second hand. For example, Hyatt, N, *Incident Investigation and Accident Prevention in the Process and Allied Industries* is offered at anything between £145 and £235 delivered (from Abe Books, October 2010).
[2] (*per* Hawkins, J., *Thrussel v. Handyside* (1888) 20 QBD at 364) quoted in Slapper and Tombs, *Corporate Crime*, p.25.
[3] Heinrich, *Industrial Accident Prevention*
[4] For summary and information on further reading, see for example
http://en.wikipedia.org/wiki/Herbert_William_Heinrich
[5] http://en.wikipedia.org/wiki/Ford_Pinto accessed October 2009: Slapper and Tombs, *Corporate Crime,* pp.119-121.
[6] Vincoli, *Basic Guide*, p.6
[7] In the UK, the government's response was to get together with the nuclear power industry to limit the damage of any bad news.
http://www.guardian.co.uk/environment/interactive/2011/jun/30/email-nuclear-uk-government-fukushima
[8] http://en.wikipedia.org/wiki/1980s#Non-natural_disasters
[9] Ridley and Channing, *Safety at Work,* p.180.
[10] See eg Reason, *Managing the Risks* Chapter 3, "Dangerous Defences"
[11] HSE, *Investigating Accidents*, p.4
[12] Manuele, *On the Practice of Safety,* p.62
[13] Sutherland, Makin and Cox, *Management of Safety*, p.9.
[14] Heinrich, *Industrial Accident Prevention*
[15] One does not have to go to extremes to find that the relationship between fatalities and major injuries, for example, is different according to the work context. According to HSE statistics for 1996/1997 there were 20 fatalities in agriculture, forestry and fishing, and only 8 fatalities in energy and water supply – but the number of major injuries was virtually the same (677 and 685 respectively). See Sutherland, Makin and Cox, *Management of Safety*, p.5.
[16] http://www.buncefieldinvestigation.gov.uk/reports/index.htm#final will give you access to the final reports. This is a good example of what a major accident investigation report looks like.
[17] Reason, *Human Error*, pp. 194-209; *Managing the Risks*, pp.1-20, 69-82
[18] http://en.wikipedia.org/wiki/Kegworth_air_disaster
[19] Reason, *Managing the Risks*, pp. 76-77; *The Human Contribution,* pp.49-51 (but note that the date of his article is wrongly printed as 1967 instead of 1987).
[20] Reason presents a number of versions of the model, the latest of which uses realistic looking slices of cheese and a mouse. Reason, *The Human Contribution,* pp.188-190
[21] http://dictionary.cambridge.org/dictionary/british/latent and under "latent" at http://oxforddictionaries.com/
[22] Manuele, *On the Practice of Safety*, p.27.

[23] Ferry, *Modern Accident Investigation*

[24] *HSG 245*, p.5.

[25] These words are in themselves worthy of psychological investigation. How do I know that someone else is wrong or silly? Psychology is a modern pursuit, and before the 1870's most of what we would now think of as psychology came within the definition of philosophy. For an introduction, see Section 1 of Radford and Gower, *A Textbook of Psychology*.

[26] If you need one, there is plenty of excellent material on the HSE website.

[27] The quoted descriptions of ego-states come from Stewart and Joines, *TA Today* p.12.

[28] I am excluding from this description cases where an ego state is excluded. For further details see Stewart and Joines, *TA Today* pp.53-55.

[29] Transactional Analysis is not, however, deterministic. If my parents are unjust to me when I am a child, I might take their behaviour in and be unjust to my own children. But I might decide, in Adult, that I will never be unjust like my parents, and behave accordingly.

[30] One of the attractions, for her, of the sulking response was that since people were often unsure what she was upset about she attracted a fair amount of attention, just as she had as a child. Addressing the problem from Adult might solve the problem, but would not get as much attention. Another "advantage" of sulking is that it punishes people, so the sulky person can feel satisfaction for "getting their own back".

[31] It dates from 1580 and probably just means playing like horses.

[32] http://www.guardian.co.uk/world/2010/sep/28/angela-merkel-stockpiling-food-east-germany accessed 24th October 2010.

[33] Controlling Parent statements often include the words "must" "should" and "have to" because Controlling Parents love to tell other people (or other ego-states) what to do. In this book I have often said things like "The committee chairman should ..." but please don't take this as Controlling Parent "should". It is conventional shorthand for "In my opinion, if we want to do this effectively, I suggest that ..."

[34] There is an interesting survey and discussion in Eiser, *Social Psychology*, Part II "Attitudes"

[35] Tilney, *Dictionary*, p.127

[36] Stewart and Joines, *TA Today* p.59.

[37] Otherwise the transactions are "crossed". If I come from Parent to Child but the other person's Child remains inaccessible to me because they are transacting from Adult, communication is not going to happen. This is quite uncomfortable so I am under psychological pressure either to break off the transactions or to change ego-state so that communication can happen.

[38] I have very much simplified this theory. For a fuller explanation see and Stewart and Joines, *TA Today* chapters 17 and 18.

[39] Stewart and Joines, *TA Today* p.178.

[40] "an overall perceptual, conceptual, affective and action set, which is used to define the self, other people and the world." *TA Today* p.329

Footnotes

---

[41] Which could lead us neatly into the theories of the drama triangle and of anger stamps - but you will have to read about those somewhere else, for example Stewart and Joines, *TA Today* chapter 21 and pp.236-238

[42] It is interesting that in the UK, ten years ago getting workers to wear PPE was a real problem as it was seen as "sissy". It was not part of the group culture. But things have changed a lot. Society has moved on, most workers' frame of reference has changed, and now wearing a high-viz jacket, boots and helmet is a matter of pride. Instead of a Parent ego-state *I'm a tough guy and I'm not wearing a helmet or boots* we have a Child ego-state *Hey, look at me! I'm a real grown- up tough construction worker in all this gear!* Getting behaviour change is often a matter of changing the group culture, and inviting contracted behaviour is one way of achieving that.

[43] Kurzman, D. (1987). *A Killing Wind: Inside Union Carbide and the Bhopal Catastrophe.* New York: McGraw-Hill. Quoted at: http://en.wikipedia.org/wiki/Bhopal_disaster

[44] Vincoli, *Basic Guide*, p.78

[45] http://en.wikipedia.org/wiki/Reactions_to_the_Deepwater_Horizon_oil_spill

[46] Detective novels use this device all the time: Mr Jones doesn't tell the police that he saw X because he knew that X *couldn't have happened.* And before long the murderer realises that Mr Jones knows something, even though Mr Jones himself doesn't know it. Mr Jones mysteriously disappears ...

[47] Laura Cameron and Tom Stocker, "Keep calm and carry on" *Safety and Health Practitioner* (March 2010, IOSH, Leicester)

[48] Harris "Directors' and engineers' responsibilities" p.8. I have paraphrased some of the questions.

[49] These include: FMEA and FMECA; Why-why analysis; Kepner-Trego; Fault tree analysis; TOPS, and even Latent Cause Analysis.

[50] Angeles, *World Class Maintenance Management*, p.209

[51] Go to www.hse.gov.uk and enter hsg245 into the search box.

[52] The accident is real but this is not from a real accident report.

[53] How likely you think an explanation is will depend on your frame of reference. I was talking to a man recently who told me that there were large numbers of spiritually aware people in a certain English city because it was built on an ancient deposit of spiritual energy which existed in the very soil and rocks of the place. There is, as far as I know (which isn't very far, I admit), no data for the number of spiritually aware people in that city, nor is there evidence that this is significantly above what would be expected by chance, nor for the existence of spiritual energy in particular locations. To him, this seemed a reasonable explanation. To me it did not. Our frames of reference were so different that the conversation could not usefully continue. On the other hand, we can sometimes be sceptical of claims and then find that there is in fact good evidence for them. A friend told me recently that she had woken up in the middle of the night to find a policeman standing beside her bed. When I suggested that this was probably a dream she was quite indignant, and later produced the policeman's friendly note advising her to make sure that, in future, she did not leave the front door wide open before she went to bed. When listening to

witness statements we need to keep an open mind; but we will require evidence for assertions and, if the evidence is not there, we will not give the assertion credibility.

[54] This is not a real accident but I am grateful to Mr Roy Bedson for supplying me with details of a similar accident and a masterly interpretation of the evidence.

[55] Vincoli, *Basic Guide*, p.1

[56] http://www.guardian.co.uk/business/2010/oct/27/glaxosmithkline-whistleblower-awarded-96m-payout

[57] Edward E Adams, *Total Quality Safety Management* (1995, American Society of Safety Engineers), p.22.

[58] Prof Kenneth Andrews, Harvard Graduate School of Business Administration, quoted in Ferry, *Modern Accident Investigation*, p.v.

[59] Vincoli, *Basic Guide*, p.12

[60] Vincoli, *Basic Guide*, p.26

[61] Krause et al, *Behaviour Based Safety Process*, p.65

[62] IET, *Safety Culture*, p.2

[63] If you want rather more ideas, there are lots of much longer descriptions eg Hughes, *Health and Safety at Work*, chapter 4

[64] You can enjoy a new hobby by spotting particularly meaningless advertising statements, which appear everywhere. My favourite is a chain of shops which puts up large signs saying "Up to 50% off!" Up to? So there could be one carefully hidden item that is reduced by 50% (reduced from what, by the way?) and everything else could be more expensive (than what)? It's meaningless; but the ulterior transaction, which no doubt many people get is *"Everything in here is 50% cheaper than in other stores"*. If a policy says it will make things "safer" we have to know what the improvement will be measured against. And when. Otherwise "safer" could mean 1% safer over the next 20 years.

[65] Neal and Griffin, "Safety Climate", p.27. "Safety climate" is not very different from "safety culture" and the differences can be ignored for our purposes.

[66] Neal and Griffin, "Safety Climate", p.29 referring to work by D. Zohar.

[67] Harris, "Directors' and engineers' responsibilities" p.9. I have paraphrased the section.

[68] Positive and negative feedback in TA is "stroke theory" but here I am going to use the behaviourist terminology because it is more familiar and provides some additional information.

[69] Most of this theory can be found in Krause et al, *Behaviour Based Safety Process*.

[70] Reason, *Managing the Risks*, p.112. On the Safety Space, see Reason, *Managing the Risks*, chapter 6; *Human Contribution*, pp.268-288.

[71] Reason, *Human Contribution*, pp.287-8

[72] Reason, *Human Contribution*, p.268

[73] Hughes and Ferrett, *Introduction to Health and Safety*, p.66

[74] ILO, *Providing safe and healthy workplaces for both women and men*, p.1

[75] 55% according to a recent report, but it has been disputed. Defenders claim the real figure is only 23%. So that's alright then.
http://www.guardian.co.uk/business/2010/nov/03/directors-pay-and-perks

Footnotes

[76] For example, HSE, *Successful health and safety management*, Chapter 5 "measuring performance"